DISASTER RISK AND VULNERABILITY

Disaster Risk and Vulnerability

Mitigation through Mobilizing Communities and Partnerships

Edited by

C. EMDAD HAQUE AND DAVID ETKIN

Published for

The Canadian Risks and Hazards Network

by

McGill-Queen's University Press
Montreal & Kingston · London · Ithaca

© McGill-Queen's University Press 2012

ISBN 978-0-7735-3963-1 (cloth)
ISBN 978-0-7735-3992-1 (paper)

Legal deposit second quarter 2012
Bibliothèque nationale du Québec

Printed in Canada on acid-free paper that is 100% ancient forest free
(100% post-consumer recycled), processed chlorine free

McGill-Queen's University Press acknowledges the support of the Canada
Council for the Arts for our publishing program. We also acknowledge
the financial support of the Government of Canada through the Canada
Book Fund for our publishing activities.

Library and Archives Canada Cataloguing in Publication

Disaster risk and vulnerability: mitigation through mobilizing
communities and partnerships / edited by C. Emdad Haque
and David Etkin.

Includes bibliographical references.
ISBN 978-0-7735-3963-1 (bound). – ISBN 978-0-7735-3992-1 (pbk.)

1. Hazard mitigation. 2. Emergency management. 3. Natural
disasters – Risk assessment. I. Haque, Chowdhury Emdadul
II. Etkin, David

HV551.2.D57 2012   363.34'6   C2011-907832-5

This book was typeset by Interscript in 10.5/13 Sabon.

Contents

Figures

Tables

Foreword

The experience of the last century and the first decade of the new millennium has revealed that the nature of disaster risk is changing in an unprecedented manner due to rapid shifts in the global economic system, the volume and range of movements of people and commodities, rapidly expanding urban density, local and regional climate and weather patterns, and above all the interface between the physical, biological, and human spheres. The human cost in lives and livelihoods resulting from nature-triggered and vulnerability-induced disasters has multiplied in recent decades. The United Nations General Assembly, observing these trends and patterns, launched the International Decade for Natural Disaster Reduction (IDNDR, 1990–99) to encourage adoption of an integrated approach to disaster risk and to launch a culture of prevention and vulnerability reduction beyond preparedness for response.

The Indian Ocean tsunami of 2004, which affected many countries; Hurricane Katrina in 2005, which devastated Louisiana and Mississippi in the United States; the Haitian earthquake, Pakistani floods, and Russian fires of 2010; and the New Zealand and Japanese earthquakes of 2011 – to cite only a few disasters – suggest that the escalating costs of such events cannot continue to be tolerated or absorbed, either by one nation alone or by the international community. The Yokohama Strategy adopted at the first World Conference on Disaster Reduction (1994) stated that all countries should aim "to defend individuals from physical injuries and traumas, protect property and contribute to ensuring progress and stability, generally recognizing that each country bears the primary responsibility for protecting its own people, infrastructure and other national assets from the impact of natural disasters, and accepting at the same time that, in the context of increasing global interdependence, concerted international cooperation and an enabling international environment are vital for the success of these national efforts." Its call for institutional cooperation and partnership at all levels was embedded in its recognition that no single community, region, or nation can prevent, mitigate, respond to, and recover quickly from a catastrophic, nature-triggered event without effective support and collaboration from partnering entities. A shared vision for a safer world and sustainable communities is grounded in the resilience of human society and the ingenuity to identify solutions to emerging kinds of risk.

Towards this goal, the member states of the United Nations adopted in 2000 a strategic framework, the International Strategy for Disaster Reduction (ISDR), to guide and coordinate the efforts of a wide range of partners to significantly reduce disaster losses and build resilient nations and communities as essential conditions for sustainable development. In 2005, the second World Conference on Disaster Reduction adopted the Hyogo Framework for Action (2005–2015): Building the Resilience of Nations and Communities to Disasters to provide further and more detailed guidance on how to reduce risk and vulnerability to natural hazards, laying the groundwork for a more effective adaptation to the increasing challenge of climate change.

This timely collection of papers explores the significance and potential of risk and vulnerability reduction through the mobilization of communities and the building and strengthening of institutional and strategic partnerships.

The revelations of insufficient risk reduction prior to the 2011 Japanese earthquake are not just about limitations in the comprehension of hazards and multiple risks at the local, sub-national, national, and international levels or failures in overall disaster risk management. They are also about failures to understand the links between physical, biological, and societal spheres and their related processes, in which the outcomes appear in the form of disaster loss and damage. Avoidance of disaster risk entirely is impossible, but human ingenuity in modifying the physical and biological systems and/or laws and social order can mitigate the impact significantly. The role of human organization and re-organization, which reflect institutions and the resilience of communities, is pivotal in advancing knowledge, planning, and implementation of development activities, disaster risk management, and policy formulation. For these and other reasons, the arguments and findings in this book should help makers of policy and practice and expand research and knowledge in various social and physical environments.

I believe that humankind, working together in partnership to promote a culture of prevention, mitigation, and resilience-building, can create a much safer, more prosperous future for coming generations.

Salvano Briceno
Senior Advisor (former Director)
UN International Strategy for Disaster Reduction (ISDR)
Geneva, Switzerland
30 March 2011

Acknowledgments

This collection of papers is the outcome of a collaborative effort that began in 1999 when we met Chris Tucker of Emergency Preparedness Canada at the annual Natural Hazards Workshop run by the Natural Hazards Research and Information Center at Boulder, Colorado, United States. Our coffee-table discussion during the lunch break triggered the First Assessment of Canadian Hazards and Disasters, which resulted in a volume on the topic and a special issue of *Natural Hazards*.

Intellectual curiosity among concerned scholars about Canada's preparedness for dealing with environmental hazards and emergencies, its efforts in the mitigation of risk and vulnerability, and its responses to environmental disaster, and a call from practitioners to know more from the scholarly literature, led to the establishment of the Canadian Risk and Hazards Network (CRHNet) in 2003. As founders of the CRHNet, we became the first co-presidents and led the organization till 2009.

The First CRHNet Symposium – "Reducing Risk through Partnerships" – took place 18–20 November 2004 in Winnipeg, and, despite career changes for both of us and our involvement in other scholarly pursuits, we never compromised the idea of publishing a collection of papers from that gathering on the role of partnership in disaster mitigation and management. This volume is the outcome of our perseverance and the writers' trust that their work would not end up unpublished, on a shelf.

This book would not have been possible without the dedication and contributions of many individuals and organizations. The First CRHNet Symposium was made possible by grants from Environment

Canada, Natural Resources Canada (Earth Science Division), the Public Health Agency of Canada, and the Natural Resources Institute of the University of Manitoba. We are grateful to Norman Halden, dean of the Clayton H. Riddell Faculty of Environment, Earth, and Resources, University of Manitoba, who, through the Clayton H. Riddell Endowment Fund, made this publication possible.

We sincerely acknowledge the invaluable work and contributions of all those who attended the 2004 CRHNet Symposium and later prepared the manuscript for publication with several iterative revisions. Many thanks to Cameron Zywina of the Office of International Relations, University of Manitoba, for his editorial support, to Mohammed Salim Uddin for preparing the final manuscript, and to Sayedur Rahman Chowdhury for his assistance in preparing the figures and maps. We would like to thank Shannon Wiebe, administrative assistant, and Tamara Keedwell, office assistant, Natural Resources Institute, University of Manitoba, for their unflagging help in administering the CHRNet Symposium and this book project. We are thankful to Salvano Briceno, Senior Advisor and former Director of the UNISDR, Geneva, for agreeing to write the foreword. Finally, we would like to thank Mark Abley, editor, and his team at McGill-Queen's University Press for their assistance with the publication of this volume.

C. Emdad Haque and David Etkin
30 March 2011

DISASTER RISK AND VULNERABILITY

Dealing with Disaster Risk and Vulnerability: People, Community, and Resilience Perspectives

C. EMDAD HAQUE AND DAVID ETKIN

Two major themes constitute the focus of this book: the significance of societal analysis for disaster research and human actions, from the perspective of integrated social-ecological systems; and the reduction of disaster risk by mobilizing communities and building partnerships among institutions. In this introduction, we assess, from a historical perspective, the magnitude of the human cost of nature-induced catastrophes; evaluate changing trends in approaches to risk, hazard, and disaster analysis; describe the context of disaster risk and vulnerability; and summarize the contents of the individual articles of this volume.

The argument that we reinforce is that without humans and their pertinent societal spheres, hazards are simply natural events and thus irrelevant; hence much attention should be paid by concerned institutions to people and communities and their capacity to engage with nature, both as a resource and as a hazard. This is particularly applicable within the context of attempts to reduce disaster risk and vulnerability. This is not a new argument, having been made by other authors (e.g., Burton, Kates, and White 1993; Wisner et al. 2004; Bankoff, Frerks, and Hilhorst 2007), but it is important to reinforce, given the continued increasing vulnerability to extreme events being exhibited by human social, economic, and technological systems.

The volume of institutional and financial resources allocated to scientific inquiries of nature- and human-triggered hazards is significant, and the increase in risk and hazards research in recent years has been marked. This is especially so since the 1980s, when the annual global loss due to catastrophic events surpassed US$100 billion, and even more since the terrorist attacks of 11 September 2001 in the United States, the Asian tsunami of 2004, Hurricane Katrina in 2005, and the Tohoku earthquake and tsunami in Japan in 2011. Increasingly, the potential effects of climate change on various natural hazards such as storms, drought, and rising sea levels and the increasing human occupancy of coastal areas are suggesting that these losses will continue to increase. Institutions at all levels in recent years have directed and redirected more financial and logistical resources to the search for scientific solutions through technological inventions and innovations, without allocating sufficient attention and resources to the assessment of the effects and effectiveness of their applications. In Kenneth Hewitt's words, such a trend is prevalent because "it accords with 'the facts' only insofar as they can be made to fit the assumptions, development and social predicaments of dominant institutions and research that has grown up serving them" (Hewitt 1983: 3). However, over the last few decades, limited but harsh criticisms of the "dominant" technocratic approach to hazards analysis have appeared in the literature (e.g., Hewitt 1983; Cannon 1994; Varley 1994; Haque 1997; Bankoff 2001; Nadeau 2006).

This book is intended to further the idea that the aspects of a community and people's power to mitigate, improve coping mechanisms, respond effectively, and recover with vigour from environmental extremes are of paramount conceptual and policy importance. Such power of people and communities is embedded in collective efforts and entities (e.g., institutions), which can be harnessed by means of mutual cooperation and partnerships. Because of the ever-increasing human and socioeconomic loss caused by environmental disasters and decreasing public funds to assist mitigation, preparedness, response, and recovery, the significance of partnerships in risk, hazard, and disaster management has risen considerably.

The papers in this book are drawn from presentations at the First CRHNet Symposium – "Reducing Risk through Partnerships" – which took place in Winnipeg, Manitoba, Canada, in November 2004, organized by the Canadian Risk and Hazards Network (CRHNet) and supported by Public Safety Canada, Natural Resources Canada, Environment Canada, and the Natural Resources Institute of the University of Manitoba. This resulting book's title – *Disaster Risk and Vulnerability: Mitigation through Mobilizing Communities and Partnerships* – reflects the contributors' shared understanding of risk and hazards as being rooted in complex social and physical systems and their associated processes. It also underscores the importance of communities, governments, non-governmental organizations, civil society organizations, and the private sector working together to avoid and reduce disaster risk and to respond more effectively.

CATASTROPHES AND HUMAN DEATHS: A HISTORICAL PERSPECTIVE

Data on catastrophic events and their impacts are difficult to gather, compile, and analyse; still, it is important to examine such information in order to put events into context. We used three sources – namely, I. Davis (2002), the Earthquake Hazards Program of the United States Geological Survey,[1] and Nash (1976) – to compile a list of catastrophic events, along with the human losses, over the past two thousand years. We added the Tohoku earthquake and tsunami of 2011 and the Asian tsunami disaster of 2004 to this list. We estimated total deaths caused by major catastrophes at approximately 157 million; this estimate excludes an additional 19 disasters that took place over the last two millennia, each of which probably killed a million or more people. In order to examine the human impact of catastrophes, we included only events with 100,000 or more estimated deaths, and the summary of the data appears in Table o.1.

It is evident that direct strikes of nature, such as major floods and earthquakes, have killed many millions of people in the last two millennia. However, it is the secondary and tertiary effects of these events that most often cascade into other socioeconomic processes, break down resource thresholds, and ultimately may cause the most human suffering. The work of De Waal (1997) and Amartya Sen's (1981) well-recognized research on famines in Ethiopia and Bangladesh, respectively, have demonstrated this point clearly, by proving that floods and droughts cause loss of employment for daily labourers. Such loss of entitlement of the poor to wage earnings and food in turn leads to starvation and famine, inducing human catastrophes. Sen proposes that the root causes of famines and epidemics are social conditions that limit the access of the poor to food and nutrition, not natural events such as floods or heat waves. De Waal (1997) echoes this in his proposition that famines occur because of the absence of social contracts between governments and their people.

Table 0.1
Natural disasters resulting in over 100,000 deaths (128 events) in the past
2,000 years

Type	Number of deaths (in millions)	Number of catastrophes with unknown deaths of a million or more
Famine	75.0	8
Diseases (total)	67.0	9
Drought	9.0	1
Flood	3.0	1
Earthquake	2.4	0
Tsunami	0.3	0

Notes:
1. The worst three disasters (influenza, plague, famine) where numerical estimates are available total
   70 million, or 44 per cent of the 159 million.
2. The tsunami of 2004 ranks about 76th on the list of 118.

Sources: L. Davis. 2002. Natural Disasters. New York: Checkmark Books; J.R. Nash. 1976. Darkest
Hours. Wallaby: Pocket Books; US Geological Survey earthquake.usgs.gov/earthquakes/world/
world_deaths.php

Disaster-related deaths are far less frequent in developed countries
than in the developing world, but the economic loss due to catastrophic events in these regions is rising astronomically (IFRCRC
2004; 2009; Munich Reinsurance Company 2009). For example, in
Canada, few people have died in comparison to outcomes in developing nations as a result of extreme events of similar magnitude and
intensity. Nevertheless, a number of disasters in Canada have caused
enormous disruption and/or damage, including the prairie droughts
of the 1930s, 1980, 1987, and 1989; Hurricane Hazel in 1954;
floods in the Saguenay in 1996 and the Red River in 1997; the ice
storm of 1998; British Columbia's forest fires of 2003; severe acute
respiratory syndrome (SARS) in 2003; Hurricane Juan in 2004; and
bovine spongiform encephalopathy (BSE). In the 1980s, Canada experienced losses of more than $1 billion from single environmental
events for the first time, and several such events have occurred since.
More important, the potential exists for disasters that would be far
more damaging than previous ones.

RISK AND HAZARDS: FROM "NATURALNESS" OF HAZARDS
TO HUMAN CAUSATION OF RISK AND HAZARDS

Traditionally, natural hazards and disasters have been treated only
through the lens of geophysical and biophysical processes, implying

that the root cause of large-scale death and destruction lies in the natural domain rather than in a coupled human–environment system. Conceptually, the physical domain has been seen as discrete and separate from human entities, and natural hazards have been defined as those elements of the physical environment harmful to humans and caused by forces extraneous to them (Burton and Kates 1964; Haque 1997). The focus on physical domains yields an incomplete understanding of natural hazards and often results in ineffective or even counterproductive solutions. Terry Cannon (1994) points out that the focus and even the title of the United Nations' International Decade for Natural Disaster Reduction (1990s) failed to move away from the "naturalness of hazards" and to emphasize the human causation of disasters. He asserted that the UN initiative, by concentrating on nature's actions, actually encouraged technical solutions to "the supposed excesses of the as yet untamed side of nature" (Cannon 1994: 17). Gilbert White and his disciples' work in the United States during the last half-century has revealed that technocratic solutions generate a false sense of safety and consequently augment human risk-taking behaviour and loss due to disaster (Kates and Burton 1986a; 1986b; Etkin 1999). Efforts during the International Decade could not therefore much affect the ever-increasing trend in disaster loss globally. None the less the experience of the increasing trend of disaster loss even in the face of remarkable technological advances and their applications has instigated communities concerned with risk and hazards at different levels (local, regional, national, and international) to question prevailing approaches and practices and to look more closely at societal dimensions. A trend to rethink the natural explanation of disasters is apparent not just in academe, but also in the practice of disaster management.

Human and societal elements are important not only because people are victims when extreme environmental events take place, but also because humans define the very essence of a "natural" hazard (see Blaikie et al. 1994; Cannon 1994; Tobin and Montz 1997; Etkin and Stevanovic 2005). Cannon (1994: 14) explains: "Nature presents humankind with a set of opportunities and risks which vary greatly in their spatial distribution. Opportunities include the many different ways in which people utilize nature for production (raw materials, energy sources) and to service their livelihoods (absorbing or recycling waste products). The risks inherent in nature consist of a wide range of hazards that put constraints on production (e.g., frosts affecting agriculture) and on other aspects of livelihoods and safety (earthquakes, floods and droughts, etc.)."

With a greater understanding emerging of the relationship of disaster processes to complex social-ecological systems, by 1980 many institutional approaches began to shift emphasis towards the role of human dimensions in hazards. For instance, the American Geological Institute (1984) redefined "natural hazard" as "a naturally occurring or man-made geologic condition or phenomenon that presents a risk or [embodies a] potential danger to life and property."

Almost three decades has passed since the publication of Kenneth Hewitt's (1983) edited collection, *Interpretations of Calamity from the Viewpoint of Human Ecology.* Still, his position statement remains challenging and merits citation. As he explains, the "dominant" paradigm in hazards and disaster research and practice is "straightforward acceptance of natural disaster as a result of 'extremes' in geophysical processes" and a technocratic view that the only way to respond is by public policy application of geophysical and engineering knowledge (Hewitt 1983: 5–7).

For Hewitt, hazards are neither explained by, nor uniquely linked with, geophysical processes that may initiate damage. This does not imply that geophysical processes are not relevant but suggests that too much causality has been attributed to them. More important, human conditions (particularly the awareness of and response to environmental hazards) do not depend solely on geophysical domains and their associated processes. Instead, hazards are more dependent on the concerns, pressures, goals, and risk-related decisions of society, not least the effectiveness of measures to mitigate calamity (Tobin and Montz 1997; Forester and Krishnan 2009). More crucially, as Hewitt shows, the causes, features, and consequences of environmental hazards and disasters cannot be fully explained by conditions and/or behaviour peculiar to catastrophic events; everyday societal forces and patterns of living play a great role. The significant elements are social order, its everyday relations to the habitat, and larger historical conditions that shape society.

In the 1990s, these perspectives were reinforced by Blaikie, Cannon, I. Davis, and Wisner (1994) with evidence from various parts of the developing world; Haque's (1997) work on floods, riverbank erosion, cyclones, and drought hazards in Bangladesh, Brazil, and Canada; and several other analysts in this field (L. Davis 1987; Wisner 1988; Mileti 1999; Pearce 2003).

During the last decade, thinking about the causation of disaster loss has shifted towards concerns about human vulnerability. Scholars have begun to distinguish between the "physical exposure" of people to threats and societal vulnerability; they have also attempted to link societal vulnerability with the propensity for hazards loss. As Fikret Berkes points out in this book, vulnerability is determined not only by exposure to hazards, but also by the resilience of the human-environment system experiencing the hazard. Terry Cannon has clarified societal vulnerability by stating that we must understand it not in terms of a given state or condition, but rather from a focus on the social, economic, political, and cultural processes that make people or society vulnerable. For him, "the vulnerability concept is a means of 'translating' known everyday processes of the economic and political separation of people into a more specific identification of those who may be at risk in hazardous environments" (Cannon 1994: 17). The argument suggests that disasters occur when an environmental hazard strikes vulnerable people. Hence, there is a link between the extent and types of vulnerability generated by people's conditions within political, social, and economic systems and the manner in which society treats hazards in terms of prevention, mitigation, preparedness, response, and recovery.

The connection of societal dimensions with climate change–induced hazards is worth noting. The Intergovernmental Panel on Climate Change's (IPCC's) Fourth Assessment (2007) emphasizes that climate change will increase the frequency of extreme weather and climatic hazards such as heat waves, El Niño events, hurricanes, droughts, and floods. Researchers on climate change adaptation now conceptualize vulnerability as a function of natural and societal processes (exposure to hazards minus adaptive capacity), where adaptive capacity is some function of environmental, social, and economic endowments. The IPCC does not formally define the term but does observe that enhancing adaptive capacity involves the promotion of sustainable development through such means as resource access, higher incomes, greater equality, and increased participation in local decision-making and actions.

From the perspectives of coupled social-ecological systems, the prevention and mitigation of hazards and disasters are possible not only by intervening in physical domains, but also (and probably more effectively) by changing and modifying societal forces, more specifically by reducing vulnerability and strengthening resilience.

Cannon's observation (1994) is just as valid today. He notes that most people concerned with disasters focus on reducing either the disaster's impact (sometimes in expensive and inappropriate ways) or human vulnerability – social protection – through certain forms of technological preparedness. Cannon (1994: 21) finds that they rarely address the major determinants that make people vulnerable (i.e., social, economic, and political factors that shape the level of resilience of people's livelihoods and their ability to withstand and prepare for hazards). Institutionally, such superficiality is still predominant (Bankoff, Frerks, and Hilhorst 2007).

Several chapters in this book discuss how resilient socio-ecological systems have the ability to learn and adjust, use all forms of knowledge, self-organize, and develop positive institutional links with other systems or sub-systems in the face of hazards. Gardner and Dekens collected evidence to support this notion from mountainous regions of India and Canada. Brenda Murphy adds that the mobilization of social capital (networks of bonds and trust) increases a community's resilience to risk and hazards. A call for a new approach to deal with risk and hazards comes from Markku Nishala, secretary general of the International Federation of Red Cross and Red Crescent Societies, who argues that, because the faces of risk and hazards are changing,

> We need new approaches that boost people's resilience to the full spectrum of physical, social and economic adversities they face. By resilience, I mean people's ability to cope with crisis and bounce back stronger than before. If we fail to shift from short-term relief to longer term support for communities in danger, we risk wasting our money and undermining the resilience we seek to enhance ... Supporting resilience means more than delivering relief or mitigating individual hazards. Local knowledge, skills, determination, livelihoods, cooperation, access to resources and representation are all vital factors enabling people to bounce back from disaster. This implies a paradigm shift in how we approach [these problems]. (IFRCRC 2004: 8–9)

Developing institutional partnerships to integrate public sectors with non-governmental organizations, the private sector, and local communities can help reduce disaster loss significantly. Recent evidence from Bangladesh reaffirms this notion (Kashem 2006; Khan and Rahman 2007; Ayers and Huque 2009).

In the face of increasingly complex relationships between human population and nature and the increasing connectedness between communities and nations, threats, risks, hazards, and disasters are changing rapidly. The depletion of natural resources, environmental degradation, the growing complexity and connectedness of human systems that depend on natural resources, marginalization, poverty, and disease are compounding nature-triggered hazards such as storms, floods, and droughts to cause chronic adversity. As well, despite unprecedented socioeconomic and technological progress, technological and cultural resources seem inadequate to deal with such compounding effects, and the result is the surpassing of thresholds and catastrophic loss.

Some recent Canadian disasters, such as the floods in the Saguenay in 1996 and the Red River in 1997, the ice storm of 1998, and the power blackout in the east in 2003, testify to that country's increasing vulnerability to catastrophic events. Even more so, other parts of the world have experienced catastrophes that also illustrate this point – for example, the Asian tsunami of December 2004, the destruction of New Orleans by Hurricane Katrina in 2005, and Japan's earthquake/tsunami of March 2011. Since we know little about emerging complex systems, we face great uncertainty about future trends and events, which seem to fit into a post-normal scientific framework.

In this context, Stephanie Chang and her colleagues offer a conceptual framework for investigating infrastructure failure interdependencies (IFIS) from the standpoint of societal impact. Their chapter clearly points to the dynamic nature of societal vulnerability, through interdependent critical infrastructure, and the need to account for these perspectives in decisions and policy-making. Also, conventional institutional measures to deal and cope with many environmental threats and hazards are proving inadequate, particularly in terms of financial and human resources. The discourse has generated a need among concerned community members to analyse and view the problems in new ways. Dennis Mileti (1999) notes that the recommendations from the second US assessment of natural hazards, intended to create a safer society, are fundamentally philosophical in nature.

THE CONTEXT OF RISK AND HAZARDS

Hazards and risk exist within a complex and changing landscape that varies in both space and time. Globally, climate change and continued environmental degradation of the natural landscape are expected to make nature-triggered disasters more frequent and severe, while social trends such as urbanization, population growth, demographic changes, globalization, the increased complexity of technological systems, and population migration to areas exposed to hazards combine to alter and increase the vulnerability of social systems in complex ways. Political decisions also affect hazards and risk. For example, in the United States after the terrorist attacks of 11 September 2001, obsession with terrorism affected how institutions and people dealt with natural, social, and other risks. Consequently, an examination of US hazard management since 2001 uncovers the reversal of several decades of progress in hazard reduction (D. Etkin 2005).

Different countries experience similar hazards in very different ways. For example, an M6.6 earthquake in the Iranian city of Bam in December 2003 killed about 26,000 people (estimates range from about 25,000 to 41,000). A similar M6.9 earthquake on 17 January 1994 in Northridge, California, resulted in 57 deaths, but also a great deal of economic damage. Buildings in Iran are not earthquake resistant, unlike those in vulnerable US regions, and are therefore subject to earthquake hazard. The 2010 earthquake in Haiti illustrates this point even more dramatically – a M7.0 earthquake 25 km west of Port-au-Prince killed perhaps 200,000 to 300,000 people. Such variation in outcome is the result not of peoples' lack of knowledge, but of differences in building codes, economic capacity, building practices, government regulation, and degree of compliance.

To the extent that data on disasters are reliable, they portray a social problem that is becoming more severe. Trends in the cost of disasters found by reinsurance companies, the Red Cross, and Public Safety Canada suggest that the problem is becoming worse. Certainly, disaster data are very difficult to interpret. For example, there exists no standard method to measure or assess impact, and comparisons are inevitably simplistic. This complexity concerning increasing costs and lack of reliable comparative indices can lead to paradoxes: society is becoming more vulnerable to extreme events even while it can better protect citizens against many of the more commonplace hazards. Even though, according to many measures (such as life span), the world is generally becoming a safer place, various authors have shown how modern social trends increase vulnerability to extremes – for example (D. Etkin 2005):

- "The fundamental contradiction in technological and economic growth … At the micro level, we find technical and economic rationality; at the macro level, technical and economic irrationality" (Vanderburg 2000).
- "We live in a world in which information … has reduced our susceptibility to accidents and diseases at the cost of increasing our vulnerability to massive social and economic catastrophes" (Slovic 2000).
- In his books, Charles Perrow argues that in complex and tightly coupled systems accidents are a fundamental property of the system, and not all of them can be prevented. Catastrophes are therefore unavoidable (Perrow 1999; 2007).
- In *Risk Society: Towards a New Modernity*, Ulrich Beck suggests that society is undergoing a transition, from one based on capital and production to one concerned mainly with risks arising in a technological society. He argues that "[a] risk society is a catastrophic society, where exceptional conditions threaten to become the norm" (Beck 1992).
- *Disasters by Design*, by Dennis Mileti, offers an overview of the second US national assessment of natural hazards (Mileti 1999). It concludes: "Too many of the accepted methods of coping with hazards have been shortsighted, postponing losses into the future rather than eliminating them." In part, this is because "people have sought to control nature and to realize the fantasy of using technology to make themselves totally safe."
- *The Ingenuity Gap* (Homer-Dixon 2000) is explicit about the adaptation deficit: "I'm convinced that if we … allow the complexity and turbulence of the systems we've created to go on increasing, unchecked – these systems will sometimes fail catastrophically … I believe this will be the central challenge – as ingenuity gaps widen the gulfs of wealth and power among us, we need imagination, metaphor and empathy more than ever, to help us remember each other's essential humanity."

These insights reflect the reality that many strategies to mitigate risk only transfer it to other groups of people or the future. Many mitigation strategies have the unintended result of engaging people and communities in taking excessive risk beyond the design standards of their infrastructure or land use planning. Such behaviour becomes so excessive that it overwhelms the risk reductions achieved in relation to more commonplace events. The results of environmental degradation, excessive risk-taking, and short-term values are a widening adaptation deficit – a gap between the risks people face and their ability to address them – and catastrophic loss of lives and properties.

The notion that major parts of risk and hazards are socially constructed assumes that the risks people choose to address and the metrics they use to represent them derive from their values and necessitate addressing the problem in an interdisciplinary manner – one that works to modify biophysical processes, to build community, and to change cultural perspectives.

DEALING WITH DISASTER RISK AND VULNERABILITY: PEOPLE, COMMUNITY, AND RESILIENCE PERSPECTIVES

A central theme of this book is the significance of societal dimensions in conceptualizing the problem of hazards and disasters and in preventing and mitigating losses from disasters more effectively. In this context, Fikret Berkes (chapter 1) extends the explanation of vulnerability beyond exposure to hazards. Since the early years of the relevant debates, Berkes has been advocating for a socio-ecological systems approach to managing natural resources and the environment. He argues that vulnerability also resides in the resilience of the system experiencing the hazard. Understanding resilience is imperative for the discussion of vulnerability for several reasons: it helps researchers assess hazards holistically in coupled human-environment systems; it stresses the system's ability to deal with a hazard, absorbing the disturbance or adapting to it; and it helps people explore policy options for dealing with uncertainty and change. As building resilience into human-environment systems is an effective way to cope with change characterized by surprises and unknowable risks, it is more central now than ever before.

Part I of this volume looks into conceptual considerations in risk and vulnerability reduction, with a focus on the role of local communities. Fikret Berkes, Brenda Murphy, Mihir Bhatt, and Tommy Reynolds, through their theoretical considerations and empirical investigations, show that resilient socio-ecological systems are capable of learning and adjusting, using all forms of knowledge, self-organizing, and developing positive institutional links with other systems or sub-systems in the face of hazards. Brenda Murphy (chapter 2) argues that communities, whether tied to particular places or not, are very central but frequently overlooked resources in both proactive and reactive phases of disaster and emergency management. Through two case studies – the 2003 electricity blackout in eastern Canada and the United States and the water disaster in Walkerton, Ontario, in 2000 – she shows that social capital (networks of strong and weak ties) can increase a community's resilience to risks and hazards and thus requires policy attention because of its crucial roles in managing hazards and emergencies.

By examining the link between disaster preparedness and response and development, Mihir Bhatt and Tommy Reynolds (chapter 3) find clear shifts in attitude among scholars, with the traditional view of disasters as one-off events giving way to recognition that development processes affect the impact of disasters. Also, within the broad context of sustainable development, disaster risk reduction must involve primary beneficiaries. The writers emphasize, by departing from mere community participation, the need for community control of initiatives that decrease susceptibility to disasters. Recent partnerships in India between communities and non-governmental organizations have shown that disaster mitigation is possible when initiatives engage the people with most to gain.

Recognizing the role of national public policies on public safety, security, and disaster and emergency management, and an enriched array of diverse policy options worldwide, part II of the book looks at international perspectives on disaster risk management and public policies. It analyses empirical work from North America (Canada) and Asia (Bangladesh, India, and Turkey). Examining Canadian public policy on disaster management, climate change, and international development, Gordon McBean (chapter 4) argues against disparate treatments of these thematic areas and for the integration of approaches. The concept of sustainable development, which guides many governments' policy framework, provides a structure for this integration. Because there is also an element of uncertainty and risk, in terms of when the next disaster will happen, how the climate will actually change, and how geopolitically critical events affect international development, risk management must play a role in addressing these emerging issues.

Gardner and Dekens (chapter 5) assert that the ability of social-ecological systems to build resilience vis-à-vis hazards is crucial to their long-term sustainability. Through examples from Canada and India, they find that resilient social-ecological systems can learn and adjust, employ all forms of knowledge, self-organize, and develop positive institutional links in the face of hazards. Their study implies that traditional social-ecological systems built resilience through avoidance, which was effective for localized hazards. Also, cross-scale institutional links appear particularly effective for building resilience in mountain social-ecological systems in the face of all contemporary hazards.

In dealing with disasters, public institutions tend to enact laws superficially, without developing full partnerships with communities at risk through social awareness and conscious self-accountability. Examining earthquakes in Turkey and its legislative initiatives, Gülkan and Karanchi (chapter 6) argue that, while observers frequently cite it as possessing strong laws vis-à-vis disaster, serious gaps remain in implementing community participation and sustained enforcement. They further insist that local communities need to develop awareness of risks and capacities for dealing with them. At the grass-roots level, assessment of disaster risks is essential and should lead to capacity building, including local institutional and community empowerment.

To examine how current emergency and disaster management strategies in the developing world have used social capital, community resilience, and other local knowledge-based resources, along with formal institutional efforts, Haque and his co-authors (chapter 7) investigated current policies and practices in Bangladesh. They reveal that, despite some strengths such as long experience in disaster response and recovery, the people's resilience, and donor agencies' support, the country has no functioning partnership among stakeholders in the formal organizational structure. This weakness flows from the lack of a culture of collective decision-making in planning, resource sharing, and managing disaster. The authors argue that a partnership framework would help the nation prevent, mitigate, prepare for, respond to, and recover from risks and hazards.

Because Canada faces a wide range of hazards and is very vulnerable to the effects of climate change, part III of this book covers the state of its environmental hazards and emergency management. We chose a vast northern nation with weather and climate extremes. It has participated in recent global increases in the costs of environmental catastrophes. Some recent disasters there, such as the 1996 Saguenay and 1997 Red River floods and the 1998 ice storm, have highlighted a disturbing trend – the marked rise in number and cost of natural disasters around the world.

Recognizing the changing environmental risks in Canada and its pioneering institutional responses, part III looks at critical aspects of disaster and emergency management in the country, including education, knowledge, perception, and action (behaviour) by various stakeholders concerning vulnerability, risk, and disasters. Bellisario, McGee, and Nirupama (chapter 8) insist that the evolving risk environment in Canada and the United States has helped transform government, stressing national security and public safety. These changes may instigate remarkable advances in emergency management if the gap between current practice and theory is acknowledged and addressed. Canada should develop education programs for emergency management that engage practitioners, policymakers, scientists, and lawmakers alongside professors. Finally, the authors call for a national agreement on standardizing curriculum in emergency management.

The effects of climate change, particularly vis-à-vis more frequent extreme environmental events, have led to intense public debate. Extreme events, such as floods, droughts, and heat waves, and their impact on individuals and communities are of special concern to policy- and decision-makers. Recognizing the importance of risk assessments and their potential mitigation measures in Canada, Parnali Dhar Chowdhury and her co-investigators (chapter 9) adopted a comparative approach to examine beliefs about floods, droughts, and heat waves arising from climate change. They mapped existing knowledge structures about hazard risks associated with climate change in rural and urban Manitoba. They developed "public models" from data they elicited from open-ended interviews with diverse community groups and "influence diagrams" from open-ended discussions to represent experts' beliefs. The study identifies knowledge gaps in six major areas and misunderstandings in two areas by juxtaposing popular and expert knowledge and perception of heat wave risk. The results suggest that the development of interactive tools for risk communication with contents more relevant to daily life could improve assessment and mitigation by both groups.

Although many residents and outsiders see Atlantic Canada as having relatively few natural hazards, the region is subject to hurricanes, storm surges, ice storms, slope failures, avalanches, extremes of heat and cold, droughts, floods, blizzards, ice storms, extreme snowfall, and earthquakes, with the greatest death toll from a single seismic event (the Burin tsunami of 18 October 1929) in Canadian history. The impacts of different natural hazards are frequently linked, increasing risk, vulnerability, and damage and human cost. Ongoing climate change could intensify many of these natural hazards, further increasing risk, vulnerability, and impact.

Within these contexts, Norm Catto (chapter 10) examined pertinent aspects and observed that the identification of natural hazards in the region, assessment of their impact, and suggestions for adaptation have proceeded in piecemeal fashion. Efforts have been hampered by the fragmentation of databases; loss of records and of institutional, community, and individual memories; limited financial and personnel resources; and difficulties with the effective dissemination of information. Comprehensive mapping and the assessment of all natural hazards are not available for any community in the region. An integrated approach, assessing risk and the vulnerability of all natural hazards in concert, is vital to reduce risk and to improve emergency planning.

In chapter 11, Stephanie Chang and her co-researchers analyse the problem of interdependent failures of critical infrastructures in disasters. Disruptions to these systems, such as electric power or transportation, frequently cause major social and economic loss in disasters, both directly and as failures in one system lead to or compound disruptions in another – infrastructure failure interdependencies (IFIs). The authors assert that strategic approaches could guide communities in preparation and mitigation. Analysing data for the ice storm of 1998, which paralysed eastern Canada, they show that that IFIs are of greatest societal concern. These infrastructure interdependencies represent potential foci for effective, targeted pre-disaster mitigation and preparedness. The framework and approach are broadly applicable across a range of natural and human-induced hazards.

CONCLUSIONS

Unless hazards or extreme events affect people, they are simply natural occurrences without social significance. Although it is bringing people into the equation that defines them, nature-triggered disasters have traditionally been treated as resulting from forces external to the human sphere. In contrast to recent social science trends, in practice hazards are still often viewed, analysed, and treated as "natural" and thus as independent entities from humans, and dealt with separately – case by case and hazard by hazard (see Tobin and Montz 1997). In cases where human intervention can change and/or modify geophysical systems, such an approach may be partially successful. However, where human awareness of, and preparedness to deal and cope with, risk and hazards are substantial, human dimensions are unavoidable aspects of hazards analysis and management. In the decades following the Second World War, even though the dominant technocratic approach allocated unprecedented financial and human resources in North America to disaster prevention, losses from nature-triggered disasters greatly increased. The research of Gilbert White and his disciples contributed greatly to this realization (Kates and Burton 1986a; 1986b).

During the last two decades, much has been written and said about the need to shift the conceptual and management approach from the technological control of geophysical forces to societal forces, where humans have more control. In order to prepare emergency/disaster managers, first responders, and communities to better deal with emerging risks and hazards, there is a need for improved public education, the design of resilient critical infrastructure, regulation concerning land use planning and zoning, and resource allocation to improve early warning. On the response and recovery side, improved collaboration and cooperation between institutions, both vertically and horizontally, have been emphasized.

A good example of such a shift has been noted after floods in the Red River Valley in Manitoba in 1997. In the United States, establishment of the federal Homeland Security Department was intended to serve a similar purpose of institutional coordination, although its incorporation and marginalization of FEMA (Federal Emergency Management Agency) have been criticized severely. However, despite the accumulation of evidence that community resilience and participatory decision-making through institutional partnerships are key to the effective management of risks and hazards, little effort is being made to bolster the role of first responders and local communities.

Given that the nature of risk and hazards is rapidly changing, knowledge institutions, community planners, emergency and disaster managers, and first responders need evolving paradigms that emphasize an adaptive approach. These institutions and people should work together to look for and provide more effective solutions. There is an increasing need to shift focus from post-disaster relief and rehabilitation to vulnerability, resilience, and mitigation. Without cross-sectoral and cross-institutional understanding and cooperation, initiatives to deal and cope with risk and hazards will remain significantly less than optimal and inefficient. In addition, top-down interventions and command-and-control approaches to large-scale emergencies and disasters are outmoded, and a participatory approach should be at the forefront for managing public safety, hazards, and disaster.

ACKNOWLEDGMENTS

The Winnipeg symposium in 2004, "Reducing Risk through Partnerships," was an initiative of the Canadian Risk and Hazards Network (CRHNet), and we would like to thank Norman Halden, dean of the Clayton H. Riddell Faculty of Environment, Earth, and Resources, University of Manitoba, for his support and advice. Public Safety Canada, Manitoba Emergency Measures Organization, Earth Science Division of Natural Resources Canada, Environment Canada, and Health Canada, co-sponsors of the symposium, made major contributions to its success and the follow-up publications. We thank Peter Hill, Irwin Itzkovitch, Chuck Sanderson, Valeriah Hwacha, Dave Hutton, and Grace Koshida for their kind assistance and contributions. Finally, this book could not have been accomplished without the help of several research and editorial assistants. We would also like to extend our special thanks to Nancy Powell Quinn, Glenn Bergen, Parnali Dhar Chowdhury, Cameron Zywina, and Mohammed Salim Uddin for their kind help and contributions to the preparation of this book.

NOTES

1 earthquake.usgs.gov/earthquakes/world/world_deaths.php

REFERENCES

American Geological Institute. 1984. *Glossary of Geology*. Falls Church, VA: American Geological Institute.

Ayers, J.M., and S. Huque. 2009. "The value of linking mitigation and adaptation: a case study of Bangladesh." *Environmental Management* 43(5): 753–64.

Bankoff, G. 2001. "Rendering the world unsafe: 'vulnerability' as Western discourse." *Disasters* 25(1): 19–35.

Bankoff, G., G. Frerks, and D. Hilhorst. 2007. *Mapping Vulnerability: Disasters, Development and People.* London: Earthscan.

Beck, U. 1992. *Risk Society: Towards a New Modernity.* Thousands Oaks, CA: Sage Publications Inc.

Blaikie, P., T. Cannon, I. Davis, and B. Wisner. 1994. *At Risk: Natural Hazards, Peoples' Vulnerability, and Disasters.* London: Routledge.

Burton, I., and R.W. Kates. 1964. "The perception of natural hazards in resource management." *Natural Resources Journal* 3: 412–14.

Burton, I., R. Kates, and G. White. 1993. *Environment as Hazard.* 2nd ed. New York: Guilford Press.

Cannon, T. 1994. "Vulnerability analysis and the explanation of 'natural' disasters." In Varley, ed., 1994: 13–30.

CRHNet (Canadian Risk and Hazards Network). 2005. D. Etkin, ed., *Reducing Risk through Partnerships: Proceedings of the 1st CRHNet Symposium.* Winnipeg: Canadian Risk and Hazards Network.

Davis, I. 1987. "Safe shelter within unsafe cities: disaster vulnerability and rapid urbanization." *Open House International* 12(3): 5–15.

Davis, L. 2002. *Natural Disasters.* New York: Checkmark Books.

De Waal, A. 1997. *Famine Crimes: Politics and the Disaster Relief Industry in Africa.* Oxford: Oxford University Press.

Etkin, D. 2005. "An expanded perspective on partnerships for the reduction of hazards and disasters." In CRHNet 2005: 37–8.

Etkin, D., and I.L. Stefanovic. 2005. "Mitigating natural disasters: the role of eco-ethics." *Mitigation and Adaptation Strategies for Global Change* 10: 469–90.

Etkin, D.A. 1999. "Risk transference and related trends: driving forces towards more mega-disasters." *Environmental Hazards* 1: 69–75.

Forester, J., and R.K. Krishnan. 2009. "Rethinking risk management policies: from 'participation' to processes of dialogue, debate and negotiation." In U.F. Paleo, ed., *Building Safer Communities. Risk Governance, Spatial Planning and Responses to Natural Hazards. NATO Science for Peace and Security Series E: Human and Societal Dynamics* 58: 34–43.

Haque, C.E. 1997. *Hazards in a Fickle Environment: Bangladesh.* Dordrecht: Kluwer Academic Press.

Hewitt, K., ed. 1983. *Interpretations of Calamity*. London: Allen and Unwin Inc.

Homer-Dixon, T. 2000. *The Ingenuity Gap. How Can We Solve the Problems of the Future?* Toronto: Alfred A. Knopf.

IFRCRC (International Federation of Red Cross and Red Crescent Societies). 2004. *World Disasters Report: Focus on Community Resilience*. Geneva: International Federation of Red Cross and Red Crescent Societies.

– 2009. *World Disasters Report: Focus on Early Warning and Action*. Geneva: International Federation of Red Cross and Red Crescent Societies.

IPCC (Intergovernmental Panel on Climate Change). 2007. Core Writing Team, R.K. Pachauri and A. Reisinger, eds. *Synthesis Report*. Contribution of Working Groups I, II and III to the Fourth Assessment Report of the Intergovernmental Panel on Climate Change. Geneva: IPCC: 104. www.ipcc.ch/publications_and_data/publications_ipcc_fourth_ assessment_report_synthesis_report.htm

Kashem, M.A. 2006. "Communications strategies for disaster preparedness in the agriculture sector in Bangladesh." *Asia-Pacific Journal of Rural Development* 16(2): 77–96.

Kates, R.W., and I. Burton, eds. 1986a. *Geography, Resources, and Environment: Volume I: Selected Writings of Gilbert F. White*. Chicago and London: University of Chicago Press.

– eds. 1986b. *Geography, Resources, and Environment: Volume II: Themes from the Work of Gilbert F. White*. Chicago and London: University of Chicago Press.

Khan, M.R., and M.A. Rahman. 2007. "Partnership approach to disaster management in Bangladesh: a critical policy assessment." *Natural Hazards* 41(2): 359–78.

Mileti, D., ed.1999. *Disasters by Design*. Washington, DC: Joseph Henry Press.

Munich Reinsurance Company. 2009. *Great Weather Catastrophes 1950– 2008*. www.munichre.com/en/ts/geo_risks/natcatservice/long-term_ statistics_since_1950/default.aspx (accessed 8 Jan. 2010).

Nadeau, R.L. 2006. *The Environmental Endgame: Mainstream Economics, Ecological Disaster and Human Survival*. Piscataway, NJ: Rutgers University Press.

Nash, J.R. 1976. *Darkest Hours*. Chicago, Illinois: Pocket Books.

Pearce, L. 2003. "Disaster management and community planning, and public participation: how to achieve sustainable hazard mitigation." In

D. Etkin, C.E. Haque, and G. Brooks, eds., *An Assessment of Natural Hazards and Disasters in Canada*. Dordrecht: Kluwer Academic Publishers: 211–28.

Perrow, C. 1999. *Normal Accidents: Living with High-Risk Technologies*. New York: Basic Books.

– 2007. *The Next Catastrophe: Reducing Our Vulnerabilities to Natural, Industrial, and Terrorist Disasters*. Princeton, NJ: Princeton University Press.

Sen, A.K. 1981. *Poverty and Famines: An Essay on Entitlement and Deprivation*. Oxford and New York: Oxford University Press.

Slovic, P. 2000. *The Perception of Risk*. London: EarthScan Publications.

Tobin, G.A., and B.E. Montz. 1997. *Natural Hazards: Explanation and Integration*. New York and London: Guilford Press.

Vanderburg, W. 2000. *The Labyrinth of Technology: A Preventive Technology and Economic Strategy as a Way Out*. Toronto: University of Toronto Press.

Varley, A., ed. 1994. *Disasters, Development and Environment*. London: John Wiley and Sons.

Wisner, B. 1988. *Power and Need in Africa: Basic Human Needs and Development Policies*. London and Trenton, NJ: Earthscan and Africa World Press.

Wisner, B., P. Blaikie, T. Cannon, and I. Davis. 2004. *At Risk: Natural Hazards, People's Vulnerability and Disasters*. 2nd ed. New York: Routledge.

PART ONE

Conceptual Considerations in Risk and Vulnerability Reduction

Understanding Uncertainty and Reducing Vulnerability: Lessons from Resilience Thinking

FIKRET BERKES

INTRODUCTION

Many studies of natural hazards focus on floods, hurricanes (cyclones), earthquakes, wildfires, ice storms, and other extreme weather events, examining why people move into disaster-prone areas and how they understand risk. Most research has taken either a physical or a human emphasis. I discuss an approach – resilience – that integrates the two and helps to explain uncertainty and reduce vulnerability. Throughout the chapter, I pursue two arguments. The first concerns the irreducible nature of uncertainty in complex systems and the necessity of living with change and uncertainty. The second relates to reducing vulnerability by building resilience.

A key concept of natural hazards studies is vulnerability – the propensity to suffer some degree of loss from a hazardous event (Etkin et al. 2004). Coppola (2007) defines vulnerability as "a measure of the propensity of an object, area, individual, group, community, country, or other entity to incur the consequence of a hazard." Turner et al. (2003) define it as the degree to which a system is likely to experience harm due to exposure to a hazard, either a perturbation (disturbance or shock) or a stress. However, it is registered not by exposure to hazards alone; it also resides in the resilience of the system experiencing the hazard (Turner et al. 2003). Resilience is a system's capacity to absorb disturbance and reorganize while undergoing change so as to retain essentially the same function, structure, identity, and feedbacks (Walker et al. 2004); there are other definitions as well (Gunderson, Allen, and Holling 2010).

Resilience is important for the discussion of vulnerability for three reasons. First, resilience thinking contributes to a comprehensive vulnerability analysis and helps provide an all-hazards approach, consistent with trends to evaluate hazards holistically (Hewitt 2004). Resilience deals with coupled human-environment systems and avoids the artificial divide between the physical and the social emphasis. Social-ecological systems tend to exhibit complex patterns and processes that are not evident when social or natural scientists analyse them separately (Liu et al. 2007).

Second, resilience emphasizes the system's ability to deal with a hazard. It allows for multiple forms of response, including the system's ability to absorb the disturbance, or to learn from it and adapt to it, or to reorganize itself. These processes often occur simultaneously ("panarchy"), across scale, in sub-systems nested in larger sub-systems (Holling 2001; Gunderson et al. 2010).

Third, because it deals with the dynamics of response to hazards, resilience is forward-looking and helps people explore policy options for dealing with uncertainty and change. As Tompkins and Adger (2004) put it, building resilience into human-environment systems is an effective way to cope with future surprises or unknowable risks. It provides a way for thinking about policies for environmental change, which is highly useful in a world of unprecedented hazards and transformations (Folke et al. 2002).

This chapter explores how resilience thinking deals with uncertainty and change in general and discusses ways to reduce vulnerability by building resilience. Examples relating to partnerships and the role of local and community approaches appear throughout. The first section deals with theory about uncertainty, complex systems, and resilience. The second looks at four aspects of resilience building: living with uncertainty, nurturing diversity, using different kinds of knowledge for learning, and creating opportunities for self-organization.

UNCERTAINTY AND RESILIENCE: THE BACKGROUND

Social and ecological systems are sufficiently complex that our knowledge of them, and our ability to predict their future dynamics, will never be complete. We must work to reduce uncertainties when possible, improve assessments of the likelihood of various important future events, and learn.

(Kinzig et al. 2000)

Practitioners know that uncertainty looms high in the area of natural hazards. Earthquakes along a major fault line are inevitable, but we cannot forecast when the next one will occur. We know that a number of hurricanes will be spawned in the tropical North Atlantic. But we cannot tell beforehand if there will be a Hurricane Mitch or Katrina among them or where exactly they will hit. We can reduce the vulnerability of Manitoba communities by creating disincentives to building on the floodplain but cannot forecast the next Red River flood of the magnitude of the 1997 one.

The importance of uncertainties is generally well known, but the irreducible nature of uncertainties in complex systems is generally not appreciated. Social and environmental systems are complex. Our knowledge of them and our ability to predict their future will never be complete, even after a great deal of research. Therefore we need a two-pronged approach. We ought to discover ways to reduce the degree of uncertainty about the dynamics of these complex systems. At the same time, we require new approaches to cope with change that we cannot predict.

Until recently, scientists in natural hazards and in many other areas usually examined biophysical factors in isolation from social factors. Yet social aspects of vulnerability are crucial and have received increasing attention in the past decade (Phillips et al. 2009). The ways in which human social systems change will depend on biophysical variables. In turn, changes in biophysical variables will depend on the extent, intensity, and type of human activity. For example, the intensity and extent of forest fires in interior British Columbia, Canada, will depend not only on ignition factors, wind, and so on, but also on the history of fire control in the area and the fuel load on the forest floor. The resilience of linked social and environmental systems will depend on the ways in which these systems have co-evolved in multi-directional feedback relationships.

We should learn when and under what circumstances a hazard or perturbation might lead to a non-linear response, out of proportion to the size of the perturbation, that might have serious and unanticipated consequences (Kinzig et al. 2000). Long-term studies of gradual change might help. But, more important, we need insights from non-linear changes, that is, when the rate of transformation suddenly alters or change occurs in a discontinuous way. Thus we require an analysis of integrated social environmental systems to improve our ability to forecast and respond to change (Berkes et al. 2003).

It would also be valuable to know when and under what circumstances a hazard might trigger a threshold effect (a breakpoint that occurs in systems with multiple stable states). Discontinuous change is often linked to crossing a threshold, although not all non-linear change is discontinuous. The shift from one stable state to another is a "regime shift" or a "flip." Such an event occurs when the threshold level of a controlling variable is exceeded, such that the nature of feedbacks changes, altering the direction (trajectory) of the system itself (Walker and Meyers 2004). Threshold effects may not be widely discussed in the hazards literature, but they are implicit, for example, in models of mudslides and avalanches. The possibility of runaway feedbacks (for example, because of permafrost thawing and methane release) crops up occasionally in the analysis of climate change (Holling 1986). Threshold effects are in fact pervasive in both biophysical systems (e.g., the breaching of a dam due to an earthquake) and social systems (e.g., a society dissolving into chaos after a war or natural disaster).

The recognition of the pervasiveness of non-linear responses and threshold effects is part of the revolution in the current science of ecology. The traditional notions of stability ("balance of nature"), linear progression, and succession that have guided ecosystem management for almost a century have given way to the idea of non-equilibrium systems, thresholds, multiple steady states, and surprises (Holling 2001; Scheffer and Carpenter 2003; Gunderson et al. 2010). Much of ecosystem research (as well as findings in other disciplines) contradicts the assumptions of predictability and controllability. Rather, ecosystems are increasingly seen as inherently unpredictable and not stable or equilibrium-centred.

Quantitative predictions of the system as a whole (as opposed to specific processes in the system) are thought to be difficult; as well, there is recognition of an irreducible unpredictability. As a result, many ecosystem-oriented ecologists have moved away from the idea of "control of nature" and "managing" ecosystems, looking instead for stable yields of products (e.g., maximum sustained yields of timber and fish). This shift has precipitated a new emphasis on the study of non-equilibrium systems and managing ecosystem processes (rather than products) for resilience (Gunderson et al. 2010).

For these ecologists, ecosystems research has moved closer to hazards research. There is a great deal of ecological work in progress dealing with shocks and stresses in ecosystems. The emphasis is on hazards that can precipitate regime shifts (flips), such as those from productive fishery systems to murky waters dominated by jellyfish, or from rich tropical forests to sun-baked tropical soils covered by scrub vegetation.

Resilience thinking, as originally developed by the Canadian ecologist C.S. (Buzz) Holling (1973; 1986), is closely related to the theory of complex adaptive systems. Because ecosystems often exhibit multiple stable states, Holling's resilience can characterize the system's capacity to maintain itself despite disturbance. Resilience theory envisions ecosystems as constantly changing and focuses on renewal and reorganization rather than on stable states. It focuses on scale (for example, sub-systems nested in one another in a panarchy) as well as on non-linear effects and thresholds. All these characteristics place resilience thinking squarely in the theory of complex adaptive systems (Holling 2001; Gunderson et al. 2010).

Holling's definition of resilience is one of many, including one in psychology that stresses the individual's ability to bounce back from adversity. Practitioners in the hazards area tend to use the concept rather loosely (e.g., Canadian Centre for Community Renewal 2000). A second definition in ecology relates to bouncing back to a reference state after a disturbance – probably less useful vis-à-vis vulnerability, since there often is no such state in coupled social and ecological systems.

Originally an ecological concept, resilience is being applied to coupled human-environment systems (or social-ecological systems). The idea is to focus not merely on ecosystems or societies per se, but on the integrated social-ecological system (Berkes and Folke 1998; Folke 2006; Berkes 2011). Many current uses of the term acknowledge reciprocal interactions between human and natural systems, underscoring the necessity to learn from past events. This development has led to consideration of the key ideas of adaptive capacity and the ability of social systems (such as institutions) to learn and adapt in response to perturbations (Folke et al. 2005).

Adaptive capacity refers to the ability of the actors in a system to influence or manage resilience. Because human actions dominate social-ecological systems, adaptability is a function mainly of the social component of the integrated system, in particular, institutions (Young 2010). Adaptation is not a mechanistic or predetermined outcome. Human agency, including the role of individuals, leaders, and institutions, is important and influences outcomes (O'Brien, Hayward, and Berkes 2009). This collective capacity to manage resilience determines whether thresholds can be successfully avoided.

In the case of climate change, for example, the production of a critical level of greenhouse gases could cause the system to exceed a threshold beyond which there may be a runaway feedback effect resulting in rapid (as opposed to gradual) global temperature change. Adaptive capacity in this case refers to society's ability to move towards the kinds of institutions, resource extraction practices, and economic organization that take advantage of new opportunities, mitigate the worst impacts, and allow for the necessary learning and innovation to cope with a new climate regime.

Organizations and institutions can "learn" as individuals do; hence, learning in the resilience sense refers to social and institutional learning, as in learning-by-doing or adaptive management (Lee 1993; Walters 1997). For the vulnerability discussion, institutional learning from society's response to previous crises and the institutions that serve as social memory is invaluable (Armitage, Berkes, and Doubleday 2007).

Institutional learning can be stored in the memory of individuals and communities. In many indigenous societies, elders hold social memory; in the industrial society, it is not clear who (if anyone) does, although much of this function may reside in various media (books, films) or storage (libraries and archives). The creation of platforms for dialogue and innovation following a crisis is key to the stimulation of learning to deal with uncertainties. It helps reorganize conceptual models and paradigms, based on a revised understanding of the conditions generating the crisis. In many cases, institutions emerge as a response to crisis and are reshaped by crisis (Folke et al. 2005).

Resilience thinking challenges the widely held notions about stability and resistance to change implicit in risk and hazard management policies around the world (Adger et al. 2005). For example, it would hold that fire prevention can increase vulnerability to large and disastrous fires, as seen in Yellowstone National Park in 1988 or in Kelowna, British Columbia, in 2003. The policy prescription of resilience would be in favour of generating disturbances (small fires) that mimic the natural fire regime in the fire-driven landscape mosaic and removing and recycling the accumulated fuel load on the forest floor. Allowing small forest fires and prescribed fires to reduce vulnerability and prevent large fires, as done in recent years, is a significantly different policy prescription from that from stability and resistance thinking.

BUILDING RESILIENCE

The synthesis chapter of *Navigating Social-Ecological Systems* (Folke, Colding, and Berkes 2003) identifies four critical factors, or clusters of factors, that interact across temporal and spatial scales and that seem to help build resilience in social-ecological systems. These factors are learning to live with change and uncertainty, nurturing diversity in its various forms, combining different types of knowledge for learning, and creating opportunity for self-organization and cross-scale links. I deal with each factor in turn, followed by a brief discussion of the obstacles to resilience-building strategies.

Living with Change and Uncertainty

Living with uncertainty starts with the observation that many long-enduring societies have developed adaptations to deal with disturbances. For example, rural communities in Bangladesh have adapted to periodic floods, as much of the country is a floodplain and agricultural and fish production depends on harnessing the flood cycle (Haque and Zaman 1993; Haque 1998). A normal flood is a "good" flood (*borsha* in the Bengali language); the abnormal one is the "bad" flood (*bonna*) (Haque 1994).

Bangladeshis are not unique in having adapted to floods. In the area of Tonle Sap, Cambodia's Great Lake, the water level fluctuates seasonally because of monsoon rains and the backflow of the Mekong River through the Tonle Sap River. In this huge floodplain – the "fish basket" of Cambodia – rural communities have three strategies to deal with water-level changes. Some migrate seasonally to higher ground; others build their houses on stilts, typically 5 to 6 metres above the ground; yet others construct and live in floating villages that bob on the lake in the high-water season (Melissa Marschke, personal communication).

How do traditional communities deal with uncertainties? Some Pacific islands have a protocol of responses to damage from major tropical cyclones. For example, when the anthropologist Raymond Firth returned to Tikopia, he found an island devastated by the cyclones; cyclones of such intensity occur on average only once every 20 years. Firth described a variety of responses: chiefs directed repair work, took measures to reduce theft, directed labour to planting rather than fishing, and sent residents abroad for wage work. Households restricted kinship obligations and reduced hospitality, ceremonies, and use of unripe crops. People used shorter fallows in agricultural fields, restricted planting and collecting rights, and demarcated boundaries more strictly than in normal times (Lees and Bates 1990).

Such responses become embedded in local institutions and stay alive in the social memory of elders. This example shows that social memory extends at least 20 years. From a hazard-management point of view, the limits of social memory are particularly interesting. After the December 2004 tsunami, several reports indicated that specialized groups of fishing peoples in Andaman Islands and Sumatra seemed to have a social memory of tsunamis and used it to avoid casualties (Adger et al. 2005). However, subsequent research on the coast of Bangladesh (A. Deb, personal communication) and in Orissa, India (M. Gadgil, personal communication), failed to detect widespread local knowledge of tsunamis. Because the most recent tsunamis in the Indian Ocean occurred in the 1940s and the 1880s (T. Murty, personal communication), this suggests that the social memory of hazards with a 60–year frequency is not common or reliable, unlike those with a 20–year cycle (Lees and Bates 1990).

Learning to live with uncertainty requires building a memory of past events, abandoning the notion of stability, expecting the unexpected, and increasing the capability to learn from crisis. "Expecting the unexpected" may seem an oxymoron, but it means having the tools and codes of conduct to fall back on when an unexpected event happens (Hewitt 2004). Major change, as in natural disasters, can of course be very damaging, but some degree of change and renewal is necessary for the system. Thus a resilient system retains the necessary elements for organization and renewal (Folke et al. 2005). Social memory (as in rules of conduct in the event of a cyclone/hurricane) and ecological memory (as in seeds that survive a forest fire) are two elements of system renewal. Each renewal cycle brings with it windows of opportunity for change.

Nurturing Diversity

The main idea underlying diversity is that it provides the seeds for new opportunities in the renewal cycle. It increases the options for coping with shocks and stresses, making the system less vulnerable. Diversification is the universal strategy for reducing risks (by spreading them out, as in an investment portfolio) and increasing options in the face of hazards (Blaikie, Wisner, and Cannon 2003; Turner et al. 2003).

At least three kinds of diversity – ecological, economic, and partnership – are relevant to the discussion of hazards. First, ecological diversity involves genetic, species, and landscape levels of biodiversity, whose availability is often crucial for resource-based rural communities. Livelihood options are in turn based on access to these resources. Many traditional societies have specialized resource-management practices and knowledge. Compared to the simplified ecosystems of agro-industrial monocultures, many traditional management systems use (and maintain) a diversity of resources that provide livelihood portfolios (Berkes and Folke 1998).

Second, the range of economic opportunities available is another aspect of diversity. As the Millennium Ecosystem Assessment (MEA) project shows, rural livelihoods and well-being depend strongly on the diversity and health of ecosystems and the services they provide, such as food, fuel, water purification, and disease regulation (MEA 2005). In regions such as northern countries, where local economies are heavily affected by large-scale resource development for the global economy, local economic diversification has emerged as a major policy objective for building resilience (Ullsten, Speth, and Chapin 2004).

Third, Mitchell (2004) emphasizes diversity of partnerships and the significance of bringing additional constituencies into the policy arena. The implications and effects of hazards, and the issues of mitigation and adaptation that they raise, generate various responses from different actors. For example, establishing tsunami warning systems is one key response at the national and international levels. At the local level, however, the main issue may be the availability of access to high land. A diversity of constituencies in the policy arena brings a range of views and considerations into the discussion, a key tenet of sustainability science (Kates et al. 2001; Turner et al. 2003). Increasing the diversity of players can lead to new thinking and expand the role of information, education, and dialogue.

Combining Different Kinds of Knowledge for Learning

There are several potential areas or ways in which science and traditional knowledge can develop collaboration and communication.

Climate change can illustrate the potential contributions of local and traditional knowledge. The combined observations from several Canadian and Alaskan Arctic indigenous communities indicate three interrelated phenomena in climate change at the local level: weather is *more variable, less predictable,* and *more frequently extreme* (Berkes 2002). These observations are consistent with international observations (IPCC 2007).

Such findings indicate that, for Inuit communities in the western Arctic, average temperature projections mean little because their change is not the major impact of climate change, which manifests itself in terms of the three observations above. Of these three, extreme weather is well known to hazard researchers studying floods, ice storms, and hurricanes (IPCC 2007). In the western Arctic, such events include extremes of warm and cold periods, extremes in snowfall, and extremes in ice freeze-up and break-up dates. Something as (apparently) simple as warm spells in midwinter can disrupt local and regional economies by interrupting transportation on ice roads, as happened in Canada's Northwest Territories in 2000–1 and in northern Manitoba in 1998, 2000, and 2002 (Berkes 2002).

These local observations and knowledge complement global science. Climate change research has resulted in numerous global circulation models. These models provide a powerful tool for research but explain little, especially at the local level. Environmental change, as a problem of complex systems, is not analysable at one level alone. Thinking about complex adaptive systems indicates that their phenomena, such as climate change, occur at multiple scales, with feedbacks across scale. Thus no single level is the "correct" one for analysis. We cannot look at climate change at the global level alone, or at the local level alone. Community-based monitoring and indigenous observations help fill in the gaps of global science and illuminate local effects and adaptations (Berkes and Folke 1998).

Bringing different kinds of knowledge together and focusing on their complementarities increase the capacity to learn. In the case of the Canadian Arctic, meetings where local experts and scientists share knowledge have generated learning environments (Berkes 2009). Such partnerships bring together parties with different relative strengths in terms of knowledge and backgrounds. For example, co-management boards administering northern land-claims agreements bring various parties together and function as bridging organizations that facilitate learning (Armitage, Berkes, and Doubleday

2007). The creation of platforms for cross-scale dialogue, allowing each partner to bring his or her expertise to the table, is particularly effective for bridging scales to stimulate learning and innovation (Cash and Moser 2000). Co-management illustrates the value of such engagement (Berkes 2009).

Creating Opportunities for Self-Organization

The resilience of a system is closely related to its capacity for self-organization because nature's cycles involve renewal and reorganization (Holling 2001). For reducing vulnerability to hazards, four aspects of self-organization merit discussion: strengthening community-based management (Berkes and Folke 1998), building cross-scale management capabilities (Folke et al. 2005), strengthening institutional memory (Folke et al. 2005), and nurturing learning organizations and adaptive co-management (Olsson, Folke, and Berkes 2004; Armitage, Berkes, and Doubleday 2007).

First, strengthening community-based management involves maintaining local capacity for social and political organization in the face of disaster. Response by the community itself, through its own institutions, is key to effective response and adaptation (Tompkins and Adger 2004). Elmqvist (personal communication) observes about his work in the Pacific that, when a recent tropical cyclone hit Samoa, the people were prepared and capable in responding (see the description above of the traditional Tikopian response). But in adjacent American Samoa, much more affluent and used to outside aid for disasters, institutions for local response were weaker and capability was largely lacking. In the Arctic, Huntington et al. (2005) point out, colonial-style administration has eroded the resilience, adaptability, and self-reliance of the traditional Inuit society, and they call for policies to strengthen communities' ability to respond to change through their own institutions.

Second, community-based management is a necessary but insufficient condition to deal with hazards in a complex world. Problems such as climate change in the Arctic show that community institutions such as Inuit hunter-trapper committees need to work with regional, national, and international organizations such as the Arctic Council. These links may be horizontal (across the same level) and/ or vertical (across levels of organization). Such governance systems can more tightly couple monitoring and response so that decisions occur close to the scene (Adger et al. 2005). The creation of systems with multi-level partnerships represents a fundamental shift from the usual top-down approach to management. It responds to the need for building resilience by using cross-scale thinking and partnerships (Cash and Moser 2000; Huntington et al. 2005).

Third, institutional memory is important for self-organization and the emergence of new, "dissipative" structures to deal with a disaster. In systems theory, these structures last only as long as the "gradient" is in place to maintain them. The equivalent in dealing with hazards might be an emergency response that allows for new communication networks, on-the-ground aid, and emergency management. This communication network probably is the most important element of an effective response and must somehow emerge out of community and regional management systems. Adger et al. (2005) note that networks set up to manage coral reefs in Trinidad and Tobago have also been invaluable in preparing for disaster.

Fourth, dynamic learning facilitates rapid innovation in the capacity to create new responses or arrangements. Such learning can be improved by adaptive co-management – a process for testing and revising institutional arrangements and environmental knowledge in a dynamic, ongoing, self-organized process of learning-by-doing (Folke et al. 2002; Berkes 2009). Learning organizations allow for errors and risk-taking as part of the learning process. Adaptive co-management combines the *dynamic learning* of adaptive management (Lee 1993) with the *links* and *partnership* of cooperative management (Armitage, Berkes, and Doubleday 2007). It typically takes place through networks of actors sharing management power and responsibility and takes the form of iterative, collaborative, feedback-based problem-solving (Olsson, Folke, and Berkes 2004).

Obstacles to Strategies for Building Resilience

A number of obstacles stand in the way of building resilience. A large literature explores planning under uncertainty, including the optimization of environmental systems under uncertainty. In water management, for example, scholars have developed "fuzzy," multi-objective optimization tools to incorporate uncertainty into decision models (e.g., Huang and Loucks 2000). Many decision-makers find it difficult to deal with the variety of considerations needed to cope with uncertainty, as the classical literature of operations research makes clear (Nacht 2004).

As for the nurture of diversity in its various forms, ecological aspects (biodiversity at various scales) are prominent in the agenda of international environmental policy, but not social, institutional, and economic aspects. Few people recognize that governance and management frameworks can help the mitigation of risk by diversifying resource use by encouraging alternative activities and livelihoods (Blaikie, Wisner, and Cannon 2003; Adger et al. 2005).

Combining types of knowledge for learning reveals that science and traditional knowledge are both essential elements of building resilience. However, integrating them into decision-making is a challenge only now provoking discussion. Experience under the Millennium Ecosystem Assessment indicates a mixed record, especially vis-à-vis resolving the power imbalance between the two knowledge systems (Reid et al. 2006).

Creating opportunity for self-organization and cross-scale links may appear a self-evident goal, but there are many complications. Self-organization is often hampered by centralized decision-making, which prevents various levels of political organization from learning from their own mistakes (Huntington et al. 2005). The building of cross-scale links has been facilitated by various multi-stakeholder processes and co-management arrangements proliferating around the world, resulting in the nurturing of learning organizations (Olsson, Folke, and Berkes 2004). Translating or communicating findings from one level to another, through boundary organizations (Cash and Moser 2000) or bridging organizations (Folke et al. 2005; Berkes 2009), appears to be a key aspect of this process.

Overcoming obstacles for resilience strategies appears to require the creation of institutions at various levels that can learn from the experience of natural hazards in an iterative way. This effort to overcome barriers involves combining different kinds of knowledge (including scientific and local), building institutions for sharing knowledge (such as bridging organizations), and generally fostering partnerships that provide complementary knowledge and skills, leading to creative problem-solving in the face of crisis and uncertainty.

CONCLUSIONS

Vulnerability is materialized by exposure to hazards, but it also resides in the resilience of the system experiencing the hazard. The concepts of vulnerability and resilience come from different

Table 1.1
Strategies very likely to enhance resilience vis-à-vis future change

Strategy	Description
Foster ecological, economic, and cultural diversity.	Diversity provides seeds for new opportunities and maximizes options for coping with change. By supporting and protecting ecological, economic, and social diversity, countries or regions make themselves less vulnerable to adverse effects of future change.
Plan for likely changes.	By recognizing the directional nature of current changes, and by identifying external drivers of change, countries may design institutional flexibility to anticipate and adjust to change.
Foster learning.	Countries, communities, NGOs, and government agencies can learn from one another. By collaborating closely to examine patterns of response to hazards, they can learn which policy options show promise. Particularly effective are learning networks of public, private, and civil society actors.
Communicate societal consequences of recent changes.	Societal consequences of hazards are felt at multiple levels. The communication of the consequences of perturbations helps people understand actual local impact and adaptation and shows that the global nature of causes warrants global action.

Source: Adapted from O. Ullsten, J.G. Speth, and F.S. Chapin (2004, 343).

scholarly traditions and literatures (Folke 2006). However, resilience is almost the flip side of vulnerability – the ability of the linked social-ecological system to deal with the hazard and make it less vulnerable. Resilience provides a conceptual tool to deal with uncertainty and future change. It is a perspective for understanding how co-evolving societies and environmental systems can cope with, and develop from, disturbance and change (Gunderson et al. 2010).

Young (2010) thinks of resilience as institutions and governance regimes' ability to handle stress in an adaptive manner. However, the application of resilience in social systems is still a work in progress. When speaking of resilience in social-ecological systems or applying the concept to governance, it is difficult to avoid clashes with key concepts in social science such as power and agency (O'Brien, Hayward, and Berkes 2009; Berkes 2011). Nevertheless, a number of policy-oriented, forward-looking international environmental assessments have incorporated resilience thinking. Examples include sustainability science (Kates et al. 2001; Turner et al. 2003), the Millennium Ecosystem Assessment (MEA 2005), and the Arctic Climate Impact Assessment (Huntington et al. 2005).

Table 1.1 reports the findings of a symposium dealing with sustainability in high latitudes. It is adapted from a document for policy-makers regarding strategies that will probably increase resilience in the face of future change (Ullsten, Speth, and Chapin 2004). With its emphasis on the diversity of options to deal with uncertainty, the value of tracking trends and flexibility, the crucial role of learning from experience, and the communication of change, Table 1.1 can be applied to many cases dealing with hazards. The importance of partnerships is underscored by the role of various players in perceiving and assessing hazards and, as the case may be, responding and adapting to them.

ACKNOWLEDGMENTS

This chapter is based on a presentation at the conference of the Canadian Risk and Hazards Network (CRHNet) in 2004 and the resulting paper in *Natural Hazards* in 2007. I thank Emdad Haque, David Etkin, and the anonymous reviewers of the 2007 paper and of this chapter for their valuable comments. This work has been supported by the Social Sciences Research Council of Canada (SSHRC) and the Canada Research Chairs program (www.chairs-chaires.gc.ca).

REFERENCES

Adger, W.N., T.P. Hughes, C. Folke, S.R. Carpenter, and J. Rockstrom. 2005. "Social-ecological resilience to coastal disasters." *Science* 309: 1036–39.

Armitage, D., F. Berkes, and N. Doubleday, eds. 2007. *Adaptive Co-Management: Collaboration, Learning, and Multi-Level Governance.* Vancouver: University of British Columbia Press.

Berkes, F. 2002. "Epilogue: making sense of Arctic environmental change?" In I. Krupnik and D. Jolly, eds., *The Earth Is Faster Now: Indigenous Observations of Arctic Environmental Change.* Fairbanks, AK: Arctic Research Consortium of the US (ARCUS): 335–49.

– 2009. "Evolution of co-management: role of knowledge generation, bridging organizations and social learning." *Journal of Environmental Management* 90: 1692–1702.

– 2011. "Restoring unity: the concept of social-ecological systems. In R. Ommer, I. Perry, P. Cury, and K. Cochrane, eds., *World Fisheries: A Social-Ecological Analysis.* Oxford: Wiley-Blackwell: 9–28.

Berkes, F., J. Colding, and C. Folke, eds. 2003. *Navigating Social-Ecological Systems: Building Resilience for Complexity and Change.* Cambridge: Cambridge University Press.

Berkes, F., and C. Folke, eds. 1998. *Linking Social and Ecological Systems: Management Practices and Social Mechanisms for Building Resilience.* Cambridge: Cambridge University Press.

Blaikie, P.M., B. Wisner, and T. Cannon. 2003. *At Risk II: Natural Hazards, People's Vulnerability and Disasters.* 2nd ed. Florence, KY: Routledge.

Canadian Centre for Community Renewal. 2000. *The Community Resilience Manual: A Resource for Rural Recovery and Renewal.* Port Alberni, BC. www. communityrenewal.ca/community-resilience-manual.

Cash, D.W., and S.C. Moser. 2000. "Linking global and local scales: designing dynamic assessment and management processes." *Global Environmental Change* 10: 109–20.

Coppola, D.P. 2007. "Preparedness." In D.P. Coppola, *Introduction to International Disaster Management.* Oxford: Elsevier: 209–40.

Etkin, D., E. Haque, L. Bellisario, and I. Burton. 2004. "An assessment of natural hazards and disasters in Canada." The Canadian Natural Hazards Assessment Project. Ottawa: Public Safety and Emergency Preparedness Canada and Environment Canada.

Folke, C. 2006. "Resilience: the emergence of a perspective for social-ecological systems analysis." *Global Environmental Change* 16: 253–67.

Folke, C., S. Carpenter, T. Elmqvist, L. Gunderson, C.S. Holling, B. Walker, et al. 2002. "Resilience and sustainable development: building adaptive capacity in a world of transformations." *International Council for Science*, ICSU Series on Science for Sustainable Development No. 3.

Folke, C., J. Colding, and F. Berkes. 2003. "Building resilience and adaptive capacity in social-ecological systems." In Berkes, Colding, and Folke (2003): 352–87.

– T. Hahn, P. Olsson, and N. Norberg. 2005. "Adaptive governance of social-ecological systems." *Annual Review of Environment and Resources* 30: 441–73.

Gunderson, L.H., C.R. Allen, and C.S. Holling. 2010. *Foundations of Ecological Resilience.* Washington, DC: Island Press.

Haque, C.E. 1994. "Flood prevention and mitigation in Bangladesh: the need for sustainable floodplain development." In R. Goodland and V. Edmundson, eds., *Environmental Assessment and Development.* Washington, DC: World Bank: 101–13.

– 1998. *Hazards in a Fickle Environment: Bangladesh.* Kluwer: Dordrecht.

- and M.Q. Zaman. 1993. "Human responses to riverine hazards in Bangladesh: a proposal for sustainable floodplain development." *World Development* 21: 93–107.

Hewitt, K. 2004. "A synthesis of the symposium and reflection on reducing risk through partnerships." Paper presented at the Conference of the Canadian Risk and Hazards Network (CRHNet), Winnipeg, MB, Nov. 18–20.

Holling, C.S. 1973. "Resilience and stability of ecological systems." *Annual Review of Ecology and Systematics* 4: 1–23.

- 1986. "The resilience of terrestrial ecosystems: local surprise and global change." In W.C. Clark and R.E. Munn, eds., *Sustainable Development of the Biosphere*. Cambridge: Cambridge University Press: 292–317.

- 2001. "Understanding the complexity of economic, ecological, and social systems." *Ecosystems* 4: 390–405.

Huang, G.H., and D.P. Loucks. 2000. "An inexact two-stage stochasitic programming model for water resources management under uncertainty." *Civil Engineering and Environmental Systems* 17: 95–118.

Huntington, H.P., S. Fox, F. Berkes, I. Krupnik, et al. 2005. "The changing Arctic: indigenous perspectives." *Arctic Climate Change Impact Assessment*. Anchorage, AK: International Arctic Science Committee of the Arctic Council: chap. 3.

IPCC (Intergovernmental Panel on Climate Change). 2007. In J.J. McCarthy, O.F. Canziani, N.A. Leary, D.J. Dokken, and K.S. White, eds., *Climate Change 2007: Impacts, Adaptation, and Vulnerability, Contribution of Working Group II to the Fourth Assessment Report of the Intergovernmental Panel on Climate Change*. Cambridge: Cambridge University Press.

Kates, R.W., W.C. Clark, R. Corell, et al. 2001. "Sustainability science." *Science* 292: 641–42.

Kinzig, A.P., J. Antle, W. Ascher, et al. 2000. *Nature and Society: An Imperative for Integrated Environmental Research*. Report prepared for the National Science Foundation of the United States.

Lee, K. 1993. *Compass and the Gyroscope*. Washington, DC: Island Press.

Lees, S.H., and D.G. Bates. 1990. "The ecology of cumulative change." In E.F. Moran, ed., *The Ecosystem Approach in Anthropology*. Ann Arbor: University of Michigan Press: 247–77.

Liu, J., T. Dietz, S.R. Carpenter, et al. 2007. "Complexity of human and natural systems." *Science* 317: 1513–16.

MEA (Millennium Ecosystem Assessment). 2005. *Ecosystems and Human Well-Being: General Synthesis*. Millennium Ecosystem Assessment. Chicago: Island Press.

Mitchell, K. 2004. "An expanded perspective on partnerships for the re-
    duction of hazards and disasters." Paper presented at the Conference of
    the Canadian Risk and Hazards Network (CRHNet), Winnipeg, MB,
    Nov. 18–20.

Nacht, M. 2004. "Operations Research." *International Encyclopedia of
    the Social and Behavioral Sciences*: 10873–6.

O'Brien, K., B. Hayward, and F. Berkes. 2009. "Rethinking social con-
    tracts: building resilience in a changing climate." *Ecology and Society*
    14 (2): 12. www.ecologyandsociety.org/vol14/iss2/art12/

Olsson, P., C. Folke, and F. Berkes. 2004. "Adaptive co-management for
    building resilience in social-ecological systems." *Environmental
    Management* 34: 75–90.

Phillips, B., D. Thomas, A. Fothergill, and L. Blinn-Pike, eds. 2009. *Social
    Vulnerability to Disasters*. Boca Raton, FL: CRC Press.

Reid, W.V., F. Berkes, T. Wilbanks, and D. Capistrano, eds. 2006. *Bridging
    Scales and Knowledge Systems*. Washington, DC: Millennium Ecosystem
    Assessment and Island Press.

Scheffer, M., and S.R. Carpenter. 2003. "Catastrophic regime shifts in
    ecosystems: linking theory to observation." *Trends in Ecology and
    Evolution* 18: 648–56.

Tompkins, E.L., and W.N. Adger. 2004. "Does adaptive management of
    natural resources enhance resilience to climate change?" *Ecology and
    Society* 9 (2): 10. www.ecologyandsociety.org/vol9/iss2/art10/

Turner, B.L., II, R.E. Kasperson, P.A. Matson, et al. 2003. "A framework
    for vulnerability analysis in sustainability science." In *Proceedings of the
    National Academy of Sciences of the United States* 100: 8074–79.

Ullsten, O., J.G. Speth, and F.S. Chapin. 2004. "Options for enhancing the
    resilience of northern countries to rapid social and environmental
    change." *Ambio* 33: 343.

Walker, B., C.S. Holling, S.R. Carpenter, and A. Kinzig. 2004. "Resilience,
    adaptability and transformability in social-ecological systems." *Ecology
    and Society* 9 (2): 5. www.ecologyandsociety.org/vol9/iss2/art5/

Walker, B., and J.A. Meyers. 2004. "Thresholds in ecological and social–
    ecological systems: a developing database." *Ecology and Society* 9 (2): 3.
    www.ecologyandsociety.org/vol9/iss2/art3/

Walters, C. 1997. "Challenges in adaptive management of riparian and
    coastal ecosystems." *Conservation Ecology* 1 (2): 1. www.consecol.org/
    vol1/iss2/art1/

Young, O.R. 2010. "Institutional dynamics: resilience, vulnerability and
    adaptation in environmental and resource regimes." *Global Environ-
    mental Change* 20: 378–85.

2

Community-Level Emergency Management: Placing Social Capital

BRENDA L. MURPHY

INTRODUCTION

Emergency management, despite the involvement of upper levels of government, tends to concentrate responsibility for the delivery of services with local authorities. It is in local spaces in most industrialized countries that planning, mitigation, and emergency preparedness take place in advance of any risk event and where response and recovery efforts occur (R.W. Perry and Nigg 1988; Mileti 1999; Canton 2007; Haddow, Bullock, and Coppola 2011). On closer examination, however, we see that what "local" entails is not at all clear. Does it equate with cities, towns, and rural spaces, or does it mean neighbourhoods and other types of social entities where people have their day-to-day lives? The former are more properly "municipalities," and the latter "communities," yet the disaster literature often conflates these terms. It seems that the way local authorities manage risk relates to, but is distinct from, the way communities do so and that municipalities and communities are not co-terminous but interact in fluid and dynamic ways.

Within the local level, therefore, there is a complex relationship between municipalities and a plethora of communities, including neighbourhoods, families, churches, service and hobby clubs, and athletic groups that have ties to that locality. For individuals and households, associational relationships within communities provide one of the mechanisms through which people organize their activities, circumscribe their identities, and muster their resources. By belonging to particular place- or interest/kinship-based communities, individuals can either increase or decrease their vulnerability to a host of potential natural, technological, and biological hazards (Barton 1969; Hulbert, Beggs, and Haines 2005).

Over the past few years an innovative body of literature called "social capital" has taken as its focus various dimensions of associational life, particularly relationships within and between communities. Social capital is distinct from financial/physical capital (in economic systems) or human capital (in individuals). The term has typically been used to explicate some of the key reasons why certain communities thrive politically, socially, and economically while others languish (Putnam 1993; Coleman 1994, Inkpen 2005). The concept is not without significant problems, particularly vis-à-vis its seeming legitimization of government cost-cutting agendas (Mohan and Mohan 2002). Social capital can also lead to both positive and negative outcomes – for instance, gangs may have strong networks of internal relationships, but this does not enhance society (Fukuyama 2001).

None the less this chapter argues that the concept can help us to understand and evaluate a community's emergency management approaches. With its concentration on relations among local actors, it allows researchers to view the local scale holistically and to comprehend some of the societal-level systemic processes that lead to either vulnerability or resilience.

While research on local emergency management has been under way for decades and some of it has focused on integrated system perspectives (e.g., Barton 1969; Bolin 1982), social capital provides a new lens through which to assess such management. As such, it highlights a range of factors and relationships that illuminate local emergency management. Further, its emphasis on the active role of community members counteracts the predominant top-down, command-and-control approach that often dominates emergency management (Dynes 2002).

This chapter examines local emergency management from the perspective of community coping capacity. Informed by the literature on community disaster, it shows how rich networks of social capital can contribute to resilience. After reviewing the literature on emergency management, communities, and social capital, it looks at two empirical examples that explore these ideas: the electricity blackout of 14 August 2003 in northeast North America and the breakout of *E. coli* water-borne disease in Walkerton, Ontario, in 2000. The first event plunged 50 million Americans and Canadians into darkness for several hours, and the second resulted in the death of seven people and the illness of 2,300 others. My attempt to study these events assessed the contribution of social capital to community emergency management.

EMERGENCY MANAGEMENT AND COMMUNITIES

Haddow et al. (2011, 1) define emergency management as "the discipline dealing with risk and risk avoidance." They maintain that it should be part of routine decision-making; it is an integral element of day-to-day security and should not be hived off into a separate disaster response silo. In their interpretation, governments from the federal to the local scale comprise the backbone of emergency management, and their book, for emergency management professionals, outlines what this emergency management entails. While most people agree that governments play a central role in emergency management, fewer, especially in practitioner-oriented literature, stress community-level emergency management. The focus of managers tends to be on command and control, as defined by emergency management personnel, which emphasizes one-way communication and the delivery of services to the public (Quarantelli 1988; Waugh and Streib 2006). This orientation does not recognize the crucial role that individuals and households play, often acting through various communities, in managing risk effectively.

Lindell and Perry (1992, 27), in a discussion of community emergency planning, express analogous sentiments. They see three key aspects to the relationship between emergency managers and community members that lead to compliance with government policies: first, emergency personnel, as government representatives, are in positions of public trust and "have the legitimate right to expect compliance" from residents; second, in emergency management, authorities are presumed to have special expertise and knowledge; and, third, community members require information and resources (Lindell and Perry 1992, 249–50). Although this approach highlights the role of authorities, the authors emphasize that, without public input, the likelihood of successful implementation diminishes greatly. They also maintain that the development of some hazards policies, in particular ones involving the delivery of service, needs the participation of affected people. Yet, they observe, in most jurisdictions residents play a minimal role in policy formation. Among other things, the engagement of community members helps determine whether emergency policies are appropriate and acceptable, encourages awareness and self-protective behaviour, and may increase the efficacy of the proposed policy. On this reading, the balance in the relationship between municipal officials and residents shifts towards a larger role for community members.

Still, many researchers suggest that the balance should shift even further. Quarantelli (1988, 56), for instance, asserts that "in good disaster planning, rather than attempting to centralize authority, it is more appropriate to develop an emergent resource coordination model. Disasters have implications for many different segments of social life and the community, each with their own pre-existing patterns of authority and each with the necessity for simultaneous action and autonomous decision making. This makes it impossible to create a centralized authority system." He asserts that disaster planning must acknowledge that disaster victims, rather than exhibiting irrational and abnormal behaviour, will maintain, to the degree possible, usual activities and family responsibilities and that organizations will continue to operate even under the most adverse conditions. The implication is that, despite authoritarian tendencies, emergency management has always involved, and will always involve, community members as active participants.

Regardless of whether the focus is on proactive disaster planning or on response, the underlying goal of emergency management is to develop strong, capable communities. Thus the rationale for community emergency management is making communities more resilient and sustainable. Resilient communities can draw on internal strengths and resources to deal with hazards and disasters (Paton and Johnston 2001). Social resilience includes community social capital, particularly the "integrating features of social organization such as trust, norms and networks" (Adger 2000, 351).

Tobin (1999, 16) delineates several characteristics of resilient and sustainable communities, including continual mitigation, efforts to reduce vulnerability for all societal members, support from responsible authorities, partnership among agencies and organizations at all levels to develop appropriate mitigative measures, and social networks strong enough "to withstand changes in the vertical and horizontal relationships through which many decisions are made." Thus resilient communities require internally strong resources and networks as well as positive links and support from their external environment.

Community Emergency Management

As we suggested in the introduction, the concepts of municipalities and communities are often conflated. For instance, Barton (1969, 49) indicates: "A direct aggregation of spatial units would give us individuals, households, *communities*, regions and nations" (emphasis added). However, in terms of a nested hierarchy of geographical scale, municipalities should properly be a middle term in governance systems, since "community" does not necessarily imply a relational positioning. It is possible to imagine a small town as a community, but it is equally plausible to talk about the emergency management community, a church community, or the global community. Municipalities are local governments, rooted in place, managing a clearly delineated territory. In Canada, municipalities are creatures of provincial governments, which give them social, political, and economic power and responsibility. In emergency management, the local approach is often mandated by provincial legislation and policies.

By comparison, the idea of community emergency management is a far more fluid concept. According to Flora (1998), communities display three characteristics: members interact on a somewhat regular basis, this interaction is not mediated by the state, and members have some degree of shared preferences or beliefs (although it may not be strong). Communities can be place-bound or interest/kinship-based groupings of people, whose boundaries may or may not coincide with politically drawn borders (Day and Murdock 1993; Swyngedouw 1997; Marston 2000).[1] Place-bound communities may consist of neighbourhoods, ethnic enclaves, areas in which functional relationships based on socioeconomic or political interaction lead to feelings of community, and so on. Interest/kinship-based communities may be tied together by kinship, affinity, or some level of common perspective. This second type of community may engage in face-to-face interactions in particular places or in relationships "stretched out" across the globe.

Similarly, Barton (1969) distinguishes between geographical units (place) and social units (interest/kinship). He maintains that both are types of social systems in which people have some interdependence and interaction. This chapter follows that line of thought but argues that the community nomenclature emphasizes the social construction of all human relationships (whether or not tied to place).

In both types of communities, membership is socially constructed – members identify their perspectives and interests as having something in common. This does not mean that communities lack form: most have some kind of internal structure that helps the group meet its needs and goals. This also does not mean that any one community can fully represent the needs and concerns of individuals and households – people tend to belong to a number of communities simultaneously, none of which individually constitutes their identity (Young 1990; McLean et al. 2002). For instance, a person might feel embedded within a neighbourhood and belong to a neighbourhood group. That same individual might belong to a kinship network, a church, a service group, an athletic club, and so on. Further, these various communities may interact formally – for example, two service clubs might join forces to raise funds – or informally, through overlapping memberships or during a search-and-rescue operation. Communities are also socially and hierarchically positioned vis-à-vis each other, with variable access to human resources, money, and political influence (Young 1990).

Under these circumstances, in contrast to the far more straightforward situation of municipalities, it is quite difficult to know what constitutes *community* emergency management or which communities may be making decisions or acting to reduce risk. Yet for local authorities responsible for emergency management, this uncertainty can result in at least two significant problems. First, since the borders of communities are rarely concurrent with municipalities, it is not always clear to what extent local governments have jurisdiction over particular communities or if risk reduction and education efforts can reach all communities within their boundaries. Second, the range of communities within a jurisdiction, compounded by the complexity of their interactions, may make it difficult to assess the vulnerability or capacity of a multitude of different, overlapping communities.

At the same time, community membership and dynamics also allow emergency management authorities to increase local resilience. First, the concurrent involvement of people in several communities, combined with the interaction of these groups, may permit the ready dissemination of information about risk management between communities. Second, since many communities have some level of internal organization – a set of norms governing behaviour, membership rules, leadership hierarchy, access to resources, and regular opportunities for interaction – it is often easy to communicate with them through these channels and to use this organizational capacity to enhance resilience or to aid in response to a risk event (Dynes 2002).

Social Capital and Community Emergency Management

Given this emergency management context, it is important to disentangle the relationships within and between communities, as well as between communities and their social milieu. In this regard, the social capital construct can provide some valuable insights (Halpern 2005). Grootaert (1998, 2) defines social capital broadly as "the set of norms, networks, and organizations through which people gain access to power and resources, and through which decision making and policy formulation occur." Woolcock (1998, 155) contends that the promises of social capital are appealing: "*Ceteris paribus*, one would expect communities blessed with high stocks of social capital to be safer, cleaner, wealthier, more literate, better governed, and generally 'happier' than those with low stocks."

Indeed, examples abound in the disaster literature of people with strong networks and relationships faring better in all phases of the hazard cycle, from planning to reconstruction. For instance, in the reconstruction efforts of voluntary organizations in Kobe, Japan, Shaw and Goda (2004) see communities with social capital making decisions collectively and thereby speeding recovery. Clason (1983), also in Japan, finds individuals in caring relationships more likely to survive a disaster – in this case, internment in civilian prisoner-of-war camps during the Second World War. In another study, Maskrey (1989) demonstrates that the key to successful mitigation is community organization and gradual, bottom-up amelioration of socioeconomic and political relationships between the state and marginalized groups. Bolin and Stanford (1998) discover that, after the earthquake in Northridge, California, community organizations were effective in helping ethnically diverse, low-income victims, in conjunction with government assistance (for a similar discussion, see Berke et al. 1993).

Despite this initial rosy picture of collaboration, several characteristics of social capital intervene to modify this image (Hooghe 2007). In Coleman (1994), social capital consists of vertical and horizontal associations within social structures that facilitate human agency.

Vertical associations are hierarchical and involve the unequal distribution of power and resources (Flora 1998; Grootaert 1998). Faupel (1987) asserts that, in the face of a disaster, these vertical relationships can funnel resources into the affected area. Several of the examples above show the value of these vertical relationships. Further, social capital is, by its nature, a public good that increases with use – it does not exist outside the network of relationships that constitute societal structure (Fukuyama 2001). As with other forms of capital, it can be gained, lost, and unevenly concentrated within different groups. Hence its presence does not in any way indicate equity within society (Grootaert 1998). In a study of vulnerability to urban flooding in Guyana, Pelling (1998) demonstrates that a weak civil society, where social capital is lacking, could not counteract the influences of ingrained political rivalries. Thus the most vulnerable and marginalized households were not able to obtain appropriate flood mitigation. Similarly, Caribbean studies reveal that groups with weaker ties to public authorities were less likely to experience timely and adequate reconstruction following a disaster (Berke et al. 1993).

For disaster research, a further corrective to the positive influence of social capital relates to the magnitude of the risk event.[2] For communities responding to disasters, the utility of social capital may depend on the magnitude of the event. That is, where a disaster obliterates most physical infrastructure and/or harms much of the community, social capital may not increase the capacity to cope. The archetypal example is the Buffalo Creek flood of 1972 in the US state of Virginia, which almost completely destroyed the valley's settlements and killed 125 people. Here, despite strong, close-knit community networks, the extent of the devastation seemed to preclude altruistic behaviour (Erikson 1976; see Erikson 1994 for other examples).

Additionally, the production and use of social capital can lead to both positive and negative externalities. Tight-knit internal cohesion and trust within a group might result in a cooperative approach to emergency management from which all members gain benefits. But strong internal social capital does not necessarily translate into between-group networking. Other communities are often constructed as outsiders and treated with suspicion or hatred (Fukuyama 2001). These concerns have been noted in the disaster literature about altruism – the way in which crises often bring out the "best within us" (Fischer 2008). Dynes and Quarantelli (1971) maintain that community conflict decreases and altruism increases under at least four conditions: when the risk event is perceived to be outside the community system; when a hierarchy of compelling immediate need emerges in the community, which in turn leads to the undertaking of urgently required activities; when overwhelming current needs reduce concerns about various social conflicts and break down established social status distinctions; and when disasters strengthen community identification through the support of primary groups and opportunities to provide community activities for the response efforts. Thus, under the right circumstances, strong social capital networks can result in positive externalities that help communities cope with disaster.

In contrast, there are several studies that associate negative externalities with social capital. Aid is not necessarily extended to strangers or people who are considered "different." For instance, aid is more likely for people with whom a relationship already exists (J.B. Perry, Hawkins, and Neal 1983), and Kaniasty and Norris (1995) conclude that in disasters a "pattern of neglect" may emerge that excludes black, older, or less educated victims from altruistic communities. Within the social capital literature, the way to overcome this situation is through the development of "weak ties" that transcend the gap between groups, often by individuals belonging to several communities (Woolcock 1998); these could emerge via opportunities for communities to interact on a regular basis.

In community emergency management, the opportunity for interaction and the development of weak ties could be part of the annual review of the municipal emergency plan. Kartez and Lindel (1987, 491) maintain that local disaster planning should involve the establishment of a "disaster assistance council" composed of all "potentially useful community organisations." This would include representatives from local volunteer and relief groups, industry, hospitals, and neighbouring jurisdictions. The authors also maintain that pre-existing groups such as Crime Watch or other neighbourhood groups could be trained to provide aid during risk events. In a similar discussion, Simpson (1992) proposes integrated disaster planning that involves government institutions, relief organizations, the private sector, and community organizations. It can be argued that municipal authorities might thereby gain access to strong social capital while promoting interaction and weak ties between communities.

One major criticism of social capital arises from its under-theorization of how "social capital can be created (and destroyed) by structural forces and institutions" (Mohan and Mohan 2002, 195). Putnam (1993), for instance, asserts that, in comparison to the south of Italy, the rich networks of relationships in the north increased prosperity. He is criticized, however, for ignoring the state's role in suppressing associations in the south. Similarly, his evaluation of the decline of social capital in the United States fails to acknowledge the federal government's role after 1945 in developing associational life and communities (Mohan and Mohan 2002). Putnam's analysis of social capital resonates with the discussion above about resilient communities, which linked resilience not only to internal resources, but also to the influence of the external context. Faupel (1987, 201), for example, in a discussion of human ecology and disasters, notes that communities are functionally and structurally related to "extra-community systems."

For liberals, a related concern is that the valuation of social capital and its benefits can be used to argue for withdrawal of state support from communities and municipalities (Skocpol 1996; Woolcock 1998). The reasoning goes that state intervention at the local scale "crowds out" community networks and may replace them with bureaucratic structures. The solution is therefore to dismantle the "welfare state," which will necessarily lead to the creation of more social capital. However, as Woolcock (1998) points out, the problem with this argument harkens back to the discussion above – states that support a stable, predictable environment and provide opportunities for a broadly participatory civil society provide the context in which "a vibrant civil society" can surface and prosper.

These are not new concerns. Bolin (1982, 256), in a discussion of long-term family recovery from disaster, observes: "Another issue of importance is the extent to which federal aid weakens 'natural' support networks and primary group aid ... If federal aid were unavailable or reduced in scope, could kin group aid and mutual support fill in the resource void?" He does not provide an answer except to suggest that, given the altruism in the response period, lower federal expenditures would be less noticeable during this time, as compared to the level of community support typically associated with recovery and reconstruction.

Dynes (2002) also notes the resonances between the disaster literature and some of the key concepts of social capital. Taking his cue from Coleman's (1994) six forms of social capital,[3] he outlines how the two perspectives intersect. Many of these points of commonality have been discussed above, but one critical point remains – the evolving role of organizations in emergency management. Dynes (2002) recaps his framework for the pattern of organizational involvement in disasters. Based on a 2x2 grid, it divides into "Tasks" (Regular and Non-regular) and "Structures" (Old and New; see Table 2.1). The resultant organizations are established (type I), expanding (type II), extending (type III), and emergent (type IV).

The work of these organizations offers specific examples of social capital in action in a disaster. *Established* organizations continue to perform the same tasks (e.g., coroner's office). *Expanding* organizations increase in size and undertake new activities (e.g., Red Cross; Tierney 1989). Both types contain groups whose goals, internal structure, and social capital networks are geared towards municipal response.

The other two types of organization evolve to assist stricken communities. *Extending* organizations take on novel tasks but keep their authority structure (e.g., a service group mobilizes to help disaster victims). *Emergent* organizations develop to meet needs perceived to be unmet by other responders. These two types exhibit both new tasks and structures (e.g., ad hoc groups formed to search and rescue, shelter victims, and so on; Tierney 1989).

The first three of the four types all involve existing organizations whose social capital can be leveraged and/or adapted in a disaster, although extending organizations often involve place- or interest/kinship-based communities. In the fourth case, emergent organizations are examples of the development of new social capital in the wake of a crisis (Dynes 2002). Depending on the circumstances, these may be ephemeral or may coalesce into full-fledged organizations over time (Drabek 1987; Drabek and McEntire 2003).

In the latter two types of organizations, the deployment and creation of social capital speak to the importance of bottom-up responses during a disaster. These same types of organizations can also contribute to emergency management during non-emergency periods (Stallings and Quarantelli 1985). This finding flies in the face of many commentators who see community members as an unorganized mass without resources or as helpless, panicky victims (Dynes 2002; Tierney 1989). This recognition helps to counter-balance top-down approaches.

Table 2.1
Organizational involvement in disasters

	Tasks	
Structures	Regular	Non-regular
Old	Established (type I)	Extending (type III)
New	Expanding (type II)	Emergent (type IV)

Source: Redrawn from R.R. Dynes. 2002. *The Importance of Social Capital in Disaster Response*. Newark: University of Delaware, Disaster Research Center, Preliminary Paper No. 327.

This literature review suggests the following definition. Community emergency management that increases resilience occurs within both place-based and interest/kinship-based communities. It addresses all aspects of the hazard cycle, from hazard identification to mitigation and preparedness through to response and recovery. It involves day-to-day activities and relationships that reduce vulnerability and increase the resilience of all members. Resilient approaches to community emergency management assume interaction within and between communities and in the relationship between communities and other social structures, such as government and patterns of social status and wealth distribution. Drawing from this discussion, the following section reviews two case studies: the 2003 power blackout and the Walkerton disaster.

TWO EXAMPLES FROM THE FIELD

Between the spring of 2001 and the winter of 2004, I worked on two case studies (Murphy 2004; Murphy et al. 2005), with part of each focusing on social capital and community emergency management.[4] This section does not recount these studies but simply demonstrates the extent to which the above approach helps explain community emergency management. To date, our study of the 2003 blackout has involved a survey of 1,203 members of the general population in Ontario in February 2004 (Figure 2.1). The purpose of the survey was to assess emergency preparedness in the wake of the event and to evaluate the impact and community response to the blackout (see Murphy 2004 for details).

The second study focuses on the *E. coli* waterborne crisis in Walkerton, Ontario (Figure 2.2), in spring 2000. Its purpose was to

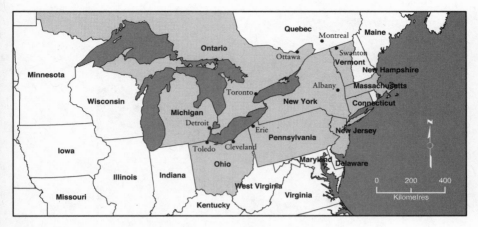

Figure 2.1
Extent of 2003 power blackout, northeastern North America

evaluate effects on the town and coping strategies by its communities.[5] Particularly in this last study, specific indicators of social capital were explored, such as membership in various groups and voluntary activity after the events.

Electricity Blackout, 14 August 2003

The blackout of 2003 affected 50 million Canadians and Americans in Ontario and several American states. Since it is rare that a risk event affects so many people simultaneously, this seemed a research opportunity that transcended the usual limitations of a case study. Among other things, our telephone survey asked respondents several questions relating to community emergency management. Since some research suggests that the nature of social capital might vary inversely with the size of the municipality (Hofferth and Iceland 1998), the survey asked respondents if they thought their neighbours would pitch in and help during an emergency. Their answers served as a rough indicator of their level of social connectedness and feelings of community. Although the overall levels of strong positive responses were quite surprising, especially among city respondents, the data support the contention that rural areas may have more robust social capital networks to access in times of crisis (Table 2.2).

Figure 2.2
Location map of Walkerton, Ontario, Canada

Other questions asked if respondents had given or received help during the blackout and the type of assistance provided. Even though the blackout caused virtually no infrastructure damage or serious injuries, 37 per cent of respondents still indicated that they provided assistance to people in need. Of those doing so, about 25 per cent indicated that they helped neighbours, just over 15 per cent, family and friends, and about 2 per cent, strangers. Conversely, only 14 per cent asserted that they had received aid. These results indicate that people used particularly their place-based networks of community social capital – based on their neighbourhoods – during the blackout. This confirms several case study findings about altruism outlined above in the literature review. Particularly important was the confirmation of altruism being offered far less to strangers; aid moved mostly through existing relationships. Also interesting was the overwhelming importance of place-based communities. Similar results were found in a study of altruism during an American blizzard (J.B. Perry, Hawkins, and Neal 1983). Since the blizzard, like the blackout, encompassed a

Table 2.2
Neighbours' helpfulness during disasters (p = .001)

	City	Town-village	Rural
Not at all likely	4%	2%	0%
Not too likely	7	5	4
Somewhat likely	26	20	14
Very likely	61	72	79
Don't know	2	1	3
Total	100%	100%	100%

widespread geographical area, it seems likely that the main community networks available early in the crises related to neighbourhoods.

Anecdotally, the blackout story in Toronto provides substance to some of these findings. At the municipal level, the power outage was considered a disaster, and this led the city to declare a state of emergency. This facilitated the deployment of the city's Emergency Operations Centre (EOC), where the efforts of fire, police, emergency medical services, and so on were efficiently coordinated to assist people caught in subways and elevators, handicapped people trapped in high-rise apartment buildings, and commuters unable to get home because public transit shut down. At the community level, stories circulated about how people coped with the disaster. For instance, the media carried many stories relating to people enjoying an evening of visiting and barbeques with neighbours, family, and friends, as well as examples of altruistic behaviour within various sorts of communities. The following is one family's story: "We were among the fortunate who had power restored shortly after midnight Friday morning. But our families in Oshawa still hadn't by early afternoon so my husband, Paul, packed up our unused Y2K generator and headed to Oshawa where he went from house to house powering up refrigerators. Late in the afternoon his parents and my parents were finally able to turn the lights on" (CBC 2003).

This example from Toronto demonstrates the interrelated but differentiated roles of municipalities and communities during disasters. Moreover, not only did communities use their social capital networks to help them cope with the event, it seems that the response also enhanced those relationships through increased opportunities for community interaction.

The Walkerton E. coli Disaster, 2000

Walkerton is a small town of 5,000 people in a rural, farming landscape. In early May 2000, one of its main wells was contaminated with E. coli and other bacteria from a nearby cattle operation. Due to the incompetence and negligence of the Public Utilities Commission (PUC), these pathogens were not properly neutralized through disinfection of the water supply. However, an exhaustive inquiry revealed that the PUC was only partly responsible for the disaster. Ongoing provincial restructuring and downloading of responsibilities onto municipalities had significantly reduced the provincial Ministry of Environment's capacity to enforce drinking water regulations, a clear reporting structure was not implemented when water testing was privatized, and the forced amalgamation of several rural areas and towns (including Walkerton) into the municipality of Brockton resulted in political and governance instability within the area.

This research project, between May 2002 and August 2003, administered a face-to-face structured survey of 104 households, undertook 23 key informant interviews with community leaders, government officials, and other people involved, and held two focus groups with 12 members of the community. The psychological and physical impact affected virtually every household in the town, although over time these effects were diminishing. About 30 per cent of households also reported financial costs. An initial round of compensation – $2,000 for each resident – offered some initial relief and satisfied most residents.[6]

However, there were many other people who were made sick by the contaminated water. It is impossible to estimate the numbers, but anecdotal evidence from residents and community leaders points to the magnitude of the problem. First, since Walkerton is the functional centre for the region, many services, places of employment, schools, and commercial operations are in the town. Second, the worst of the contamination occurred on a warm, early-summer holiday weekend. Many residents had family and friends visiting from other areas, and there was at least one wedding and a ball tournament that weekend. Many of these outsiders also became ill drinking local water. Thus members of both place- and interest/kinship-based communities experienced illness related to the *E. coli* contamination. Many of these outsiders struggled to receive compensation or were never able to obtain funding.

Within Walkerton, there exists an extensive set of social capital networks. The survey of residents revealed that 72 per cent were involved in some type of community activity. Of those people, about one-third indicated that they relied on these organizations for assistance during the crisis. Furthermore, 34 per cent of respondents said that their volunteer activity during the crisis related to their prior involvement with community groups. This indicates that the established, formal networks formed around community groups provided a good conduit for people to help their fellow residents during the crisis. Apart from these organized groups, 60 per cent of respondents helped their neighbours in some way. In terms of social capital, 63 per cent stated that there was either quite a bit of or an extreme reliance on close ties with family and friends (strong social capital), while reliance on more distant family and friends was rated at 32 per cent (weak social capital). Also evaluated were the statistical associations between the levels of recovery and participation in community organizations; between recovery and reliance on organizations; and between recovery and reliance on family and friends. These were all generally found to be positively associated. These results point to the contribution of social capital networks to the development of resilient communities.

At the outset of the Walkerton crisis, the unstable political climate within the new municipality of Brockton,[7] concern over the assignment of blame, plus fears about provincial appropriation of authority, meant that a state of emergency was not declared and that the initial leadership response was uncertain. As a result, the normal disaster response mechanisms, structured by an emergency operations centre (EOC), were not activated. Certainly emergency medical services responded, but other typical first responders (type I organizations), such as fire and police, were underused, and at the outset there was little coordination of activities or provision of response services. For residents, the main infrastructure problem related to the "boil-water" order that left them with no readily available source of potable water. Bottled water was not initially supplied by the province or any other authority. Instead, donations began pouring in from all over Canada.

In a stellar demonstration of the value of social capital, as transport trucks began to converge on Walkerton, with no EOC or other municipal authority to take charge, local residents took the lead. It was decided that the water should be dropped off at the local arena, with the unloading, organization, and eventual distribution of the water coordinated by an anarchic blend of local municipal employees, available fork-lift operators, business owners, service clubs, and virtually anyone else who wanted to assist.

Someone posted a sign-up sheet on the wall. Over the following few weeks, until a provincially appointed manager was finally hired in July, community members developed a sophisticated water-distribution network. Some people worked within the arena sorting deliveries, and others loaded bottles into vehicles as residents lined up to obtain their water. Still other residents pressed available cube vans and similar vehicles into service to deliver water to people not able to go to the distribution centre. When asked, the interviewees could not pinpoint how the system began, how needs were identified, or how decisions were made – this was all handled on an ad hoc basis. This response is an excellent example of the development of Dynes's type-IV, emergent organizational structure to help residents. It also resulted in new social capital relationships.

In this case, this new organization did not develop in a vacuum. It was clear that the close-knit relationships in Walkerton were the basis for it to emerge. Existing organizations contributed to the water-distribution centre (e.g., undertook new tasks) through their established structures and resources (type III, extending organizations). For instance, often local groups would sign up en masse to help at the water-distribution centre. A service group or athletic team would block out several shifts on the sign-up sheet, and then members would organize among themselves who would be available and at what time. Thus the centre simultaneously involved both extending *and* emergent organizational structures, seemingly leaping the divide between weak and strong social capital.

Beyond water distribution, two other emergent organizations developed in this crisis. The Walkerton Community Foundation (WCF) was formed to handle some of the money that was donated to Walkerton. Echoing the organizational structure that developed for the water-distribution centre, the WCF consists of representatives from several established organizations, including service clubs. However, due to the varied (and occasionally conflicting) mandates of these organizations, there were several disagreements regarding how the money should be spent and later if the WCF should present a unified group perspective at the inquiry. This representation of the WCF as a community with a common perspective was deeply troubling to some members, and eventually a few withdrew. This is a good example of the tension between weak and strong capital. In this case, strong internal social capital made it difficult to bridge the gap and contribute to this new organization.

The final emergent organization was Concerned Walkerton Citizens (CWC), which focused on political lobbying. Its efforts have been quite successful – the Walkerton Inquiry was held in the town and produced a two-volume report with an extensive set of recommendations that are still being implemented. This is an example of an emergent group that solidified into a stable organization with new social capital networks.

As in other disasters elsewhere, many existing organizations used their social capital to help deal with the crisis. In contrast to the problems within WCF, for instance, Kinsmen/Kinettes, with little internal acrimony, distributed a second, substantial pot of money. Beyond the work at the water-distribution centre, extending (type III) organizations, such as the Knights of Columbus and the Rotary Club, delivered flyers, collected donations, and helped vulnerable community members. In their response efforts, these organizations used their horizontal social capital relating to their local members and resources. As well, they used their vertical relationships with higher-level group authorities (e.g., national offices) and their horizontal relationships with other chapters of their organization to further their contribution. For example, when many sick Walkerton residents were transported out to a large city hospital, two hours away, groups such as the Knights of Columbus called on similar organizations in that city to arrange for the housing of relatives. Many groups also used their umbrella organizations to obtain sizeable donations for the town.

In other cases, group characteristics actually contributed to vulnerability rather than to resilience. For instance, the Legion did not participate actively in the response because so many of its members were elderly and therefore quite susceptible to the waterborne contamination. Many members became quite sick and needed assistance.

Particularly in the delivery of social services, there were also several examples of expanding (type II) organizations. Home and

Community Support Services of Grey-Bruce[8] and some of its affili-
ates, such as Meals on Wheels, provide a range of valuable services
such as transportation to medical appointments and meals for sick
and elderly people. These groups assist some of the most vulnerable
members of the community. During and after the crisis, in addition
to their regular clients, they also provided services to people made ill
by the tainted water. They also tried to provide supplementary services such as making sure clients understood the boil-water advisory
and arranging for supplies of water to be delivered. Unfortunately,
during the crisis these types of groups were not treated as important
responders. Instead, they struggled to obtain information to help
their clients. This situation points to the need for a more sophisticated understanding of the players in community emergency management, since their existing social capital networks contribute
greatly to community resilience.

CONCLUSIONS

This analysis of the two case studies highlights the way in which the
social capital embedded or emerging within various organizational
structures, as well as within kinship, friendship, and neighbourhood
relationships, helps make communities more resilient. In other words,
both place-based and interest/kinship-based communities are impli-
cated in the development of resilience. This complex interweaving of
bonds and ties forms the fabric of resilience; it also suggests that vulnerability is heightened for people outside this safety net. Additionally,
communities have their own internal vertical and horizontal positioning, and they are juxtaposed in relation to each other. Depending
on the context, this affects their capacity to manage an emergency.
These types of scalar relationships extend to community interactions
with government authorities, who often elide political and community borders, may or may not provide leadership during crisis, and
may undercut local resilience through cost-cutting and restructuring.
Thus the existence of social capital does not guarantee resilience.
Both communities and municipal emergency management must try
to identify and foster the social capital bonds that contribute to resilience and seek out marginalized people to reduce their vulnerability.
These are key mechanisms for achieving resilience for everyone.

At the municipal level, the difference between the two case studies
also highlights the way in which municipal emergency management
can affect community resilience. In the case of the blackout, since

many municipal governments (including Toronto's) responded formally to the event, the community response provided an additional source of support for local people. In contrast, in Walkerton, where the formal municipal response was less coordinated and more uncertain, the community essentially filled a void. While residents coped admirably, the local response might have been more robust, and the inadequate municipal approach undermined resilience.

The focus here on social capital, resilience, and community emergency management can be seen as a shift from the traditional approaches regarding the understanding of how communities, in contradistinction to municipalities, prepare for, and respond to, crisis. This orientation distinguishes between the "official" emergency management structure, organized through the municipality, and the more fluid, contingent emergency management approach that exists within communities. This does not imply a downgrading of the municipal role; rather it suggests that resilience requires community involvement *in addition to* official activities. The shift towards recognition of community actors acknowledges their inherent capacities and resources and seeks to improve these abilities. In this respect, municipal emergency managers can provide opportunities for residents to learn more about hazards in their locality and foster the social capital bonds that contribute to resilience. Whether through a review of the emergency plan, through other municipal initiatives, or through cultural and sports events, opportunities to develop involved, interactive communities may ultimately increase their resilience.

ACKNOWLEDGMENTS

The author gratefully acknowledges the financial support of the Brantford, Ontario, campus of Wilfrid Laurier University and the Institute for Catastrophic Loss Reduction at the University of Western Ontario in London. The author also wishes to acknowledge helpful comments on an earlier version of this manuscript by C. Emdad Haque and David Etkin.

NOTES

1 This characterisation separates place-based communities from others to emphasize their geographical differences. Kinship- and interest-based communities could be further differentiated, but this is beyond the scope of this chapter.

2 Magnitude is used here as a general term to denote several disaster characteristics, including rate of onset and duration. As David Etkin indicates, a
   community's response, either positive (cooperative) or negative (competi-
   tive), may also be influenced by several contexts or may shift over time.
   For instance, communities initially cooperative after a disaster may become less cohesive if recovery is slow or unresponsive to their needs. These
   ideas remain to be elaborated later.

3 According to Dynes (2002, 5), Coleman's (1994) six forms include "obligations and expectations, informational potential, norm and effective
   sanctions, authority relations, appropriable social organizations, and in-
   tentional organizations."

4 A third study, on the tornado in Pine Lake, Alberta, in 2000 also assesses
   some aspects of social capital. See Murphy et al. (2005).

5 Full details of the Walkerton study available from the author s; Murphy
   and Dolan, unpub. ms.

6 A second round of compensation was also initiated. The significant problems associated with this are outlined in Murphy and Dolan, unpublished
   manuscript.

7 At least for Walkerton, one of the unintended consequences of the restructuring of local space was the disruption of established social capital networks. It appears that these networks can adjust over time, but until then
   places such as Brockton tend to be more vulnerable and less able to deal
   efficiently with adverse events.

8 Grey-Bruce is the double county served by the support services.

REFERENCES

Adger, W.N. 2000. "Social and ecological resilience: are they related?" *Progress in Human Geography* 24: 347–64.

Barton, A.H. 1969. *Communities in Disaster*. Garden City, NY: Doubleday
   and Company Inc.

Berke, P., J. Kartez, and D. Wenger. 1993. "Recovery after disaster: achieving sustainable development, mitigation and equity." *Disasters* 17:
   93–109.

Bolin, R. 1982. *Long-Term Family Recovery from Disaster*. Boulder:
   University of Colorado.

– and L. Stanford. 1998. "The Northridge earthquake: community-based
   approaches to unmet recovery needs." *Disasters* 22: 21–38.

Canton, L.G. 2007. *Emergency Management: Concepts and Strategies for
   Effective Programs*. Hoboken, NJ: Wiley Inter-Science.

CBC (Canadian Broadcasting Corporation). 2003. Personal Stories. CBC News Online, 15 Aug. www.cbc.ca/news/background/poweroutage/personalstories.html

Clason, C. 1983. "Family as lifesaver in disaster." *International Journal of Mass Emergencies and Disasters* 1: 43–62.

Coleman, J. 1994. *Foundations of Social Theory.* Cambridge, MA: Belknap Press.

Day, G., and J. Murdoch. 1993. "Locality and community: coming to terms with place." *Sociological Review* 40: 82–111.

Drabek, T., and D. McEntire. 2003. "Emergent phenomena and the sociology of disaster: lessons, trends and opportunities from the research literature." *Disaster Prevention and Management* 12(2): 97–113.

Drabek, T.E. 1987. "Emergent structures." In R.R. Dynes et al., eds., *Sociology of Disasters: Contribution of Sociology to Disaster Research.* Milan: Franco Angeli Libri s.r.l: 259–90.

Dynes, R.R. 2002. *The Importance of Social Capital in Disaster Response.* Newark: University of Delaware, Disaster Research Centre, Preliminary Paper No. 327.

Dynes, R.R., and E.L. Quarantelli. 1971. "The absence of community conflict in the early phases of natural disasters." In G. Glagett and G. Smith, eds., *Conflict Resolution: Contributions of the Behavioral Sciences.* Notre Dame, IN: University of Notre Dame Press: 200–4.

Erikson, K. 1976. *Everything in Its Path.* New York: Simon and Schuster.

– 1994. *A New Species of Trouble: Explorations in Disaster, Trauma, and Community.* New York: W.W. Norton and Company.

Faupel, C.E. 1987. "Human ecology and disaster: contributions to research and policy formation." In R.R. Dynes et al., *Sociology of Disasters: Contribution of Sociology to Disaster Research.* Milan: Franco Angeli Libri s.r.l: 182–211.

Fischer, H.W. 2008. *Response to Disaster: Fact versus Fiction and Its Perpetuation.* 3rd ed. New York: University Press of America.

Flora, J.L. 1998. "Social capital and communities of place." *Rural Sociology* 63: 481–506.

Fukuyama, R. 2001. "Social capital, civil society and development." *Third World Quarterly* 22: 7–20.

Grootaert, C. 1998. "Social capital: the missing link?" The World Bank, Social Development Family, Environmentally and Socially Sustainable Development Network, Social Capital Initiative. Working Paper No. 3. Washington, DC.

Haddow, G.D., J.A. Bullock, and D.P. Coppola. 2011. *Introduction to Emergency Management*. 4th ed. Burlington, MA: Butterworth-Heinemann.

Halpern, D. 2005. *Social Capital*. Cambridge, England, and Malden, MA: Polity.

Hofferth, S.L., and J. Iceland. 1998. "Social capital in rural and urban communities." *Rural Sociology* 63: 574–98.

Hooghe, M. 2007. "Social capital and diversity generalized trust, social cohesion and regimes of diversity." *Canadian Journal of Political Science* 40: 709–32.

Hulbert, J.S., J.J. Beggs, and V.A. Haines. 2005. "Social networks and social capital in extreme environments." In N. Lin, K. Cook, and R.S. Burt, eds., *Social Capital: Theory and Research*. New Brunswick, NJ: Aldine Transaction: 209–32.

Inkpen, A. 2005. "Social capital, networks and knowledge transfer." *Academy of Management Review* 30(1): 146–65.

Kaniasty, K., and F.H. Norris. 1995. "In search of altruistic community: patterns of social support mobilization following Hurricane Hugo." *American Journal of Community Psychology* 23: 447–77.

Kartez, J.D., and M.K. Lindell. 1987. "Planning for uncertainty: the case of local disaster planning." *APA Journal* autumn: 487–98.

Lindell, M.K., and R.W. Perry. 1992. *Behavioral Foundations of Community Emergency Planning*. Washington, DC: Hemisphere Publishing Corporation.

Martson, S.A. 2000. "The social construction of scale." *Progress in Human Geography* 24: 219–42.

Maskrey, A. 1989. *Disaster Mitigation: A Community Based Approach*. Oxford: Oxfam Print Unit.

McLean, S.L., D.A. Schultz, and M.B. Steger. 2002. *Social Capital: Critical Perspectives on Community and "Bowling Alone."* New York: New York University Press.

Mileti, D.S. 1999. *Disaster by Design: A Reassessment of Natural Hazards in the United States*. Washington, DC: Joseph Henry Press.

Mohan, G., and J. Mohan. 2002. "Placing social capital." *Progress in Human Geography* 26: 191–210.

Murphy, B.L. 2004. "Emergency Management and August 14th, 2003, Blackout." ICLR Research, Paper Series No. 40. Institute for Catastrophic Loss Reduction, University of Western Ontario, London, ON. www.iclr.org/pdf/Emergency%20Preparedness%20and%20the%20blackout2.pdf

Murphy, B.L., L. Falkiner, G. McBean, A.H. Dolan, and P. Kovacs. 2005. *Enhancing Local Level Emergency Management: The Influence of Disaster Experience and the Role of Households and Neighbourhoods.* Report for the Institute of Catastrophic Loss Reduction. London, ON: University of Western Ontario.

Paton, D., and D. Johnston. 2001. "Disasters and communities: vulnerability, resilience and preparedness." *Disaster Prevention and Management* 10: 270–7.

Pelling, M. 1998. "Participation, social capital and vulnerability to urban flooding in Guyana." *Journal of International Development* 10: 469–86.

Perry, J.B., R. Hawkins, and D.M. Neal. 1983. "Giving and receiving aid." *International Journal of Mass Emergencies and Disasters* 1: 171–88.

Perry, R.W., and J.M. Nigg. 1988. "Emergency preparedness and response planning: an intergovernmental perspective." In M. Lystad, ed., *Mental Health Responses to Mass Emergencies.* New York: Brunner/Mazel, Inc.: 346–70.

Putnam, R. 1993. *Making Democracy Work.* Princeton, NJ: Princeton University Press.

Quarantelli, E.L. 1988. "Assessing disaster preparedness planning: a set of criteria and their applicability to developing countries." *Regional Development Dialogue* 9: 48–69.

Shaw, R., and K. Goda. 2004. "From disaster to sustainable civil society: the Kobe experience." *Disasters* 28: 16–40.

Simpson, D.M. 1992. "Risk and disaster: arguments for a community-based planning approach." *Berkley Planning Journal* 7: 98–120.

Skocpol, T. 1996. "Unravelling from above." *American Prospect* 25: 20–5.

Stallings, R.A., and E.L. Quarantelli. 1985. "Emergent citizen groups and emergency management." *Public Administration Review* 45: 93–100.

Swyngedouw, E. 1997. "Neither global [n]or local: 'globalization' and the politics of scale." In K. Cox, ed., *Spaces of Globalization: Reasserting the Power of the Local.* New York: Guilford Press: 137–66.

Tierney, K.J. 1989. "The social and community contexts of disaster." In R. Gist and B. Lubin, eds., *Psychosocial Aspects of Disaster.* Toronto: John Wiley and Sons, Inc.: 11–39.

Tobin, G.A. 1999. "Sustainability and community resilience: the holy grail of hazards planning?" *Environmental Hazards* 1: 13–25.

Waugh, W., and G. Streib. 2006. "Collaboration and leadership for effective emergency management." *Public Administration Review* 66: 131–40.

Woolcock, M. 1998. "Social capital and economic development: towards a theoretical synthesis and policy framework." *Theory and Society* 27: 151–208.

Young, I.M. 1990. *Justice and the Politics of Difference*. Princeton, NJ: Princeton University Press.

3

Community-Based Disaster Risk Reduction: Realizing the Primacy of Community

MIHIR R. BHATT AND TOMMY REYNOLDS

INTRODUCTION: PARADIGM SHIFT
FROM RELIEF TO PREVENTION

Development and Disasters

From 1987 to 1989, communities in rural Gujarat, India, experienced a prolonged drought. The crisis decreased the savings of poor families, reduced food stockpiles, and damaged soil. The combined impact left agricultural workers and consumers of their produce more vulnerable; it depleted savings and decreased purchasing power. A task force that later became the All India Disaster Mitigation Institute (AIDMI) studied the impact of the drought on local development and found progress impeded. The disaster derailed hard-won development gains by diverting precious resources from long-term goals. At the household level, assets diverted might instead have been saved to purchase shelter, build up livelihood capital, or gain access to improved water sources. At the national level, such diverted assets might have covered health care or education; instead they went to humanitarian assistance.

The United Nations Development Programme, or UNDP (2005), cites two other examples of disasters with similar impacts: the 2004 hurricane in Haiti not only killed 2,000 people but also reduced economic growth by 1 per cent that year; hurricanes in Grenada in 2004 caused a similar drop in economic growth (UNDP 2005).

During the 1990s and the first decade of this century, a variety of individuals, practitioners, academics, and organizations stressed the correlation between hazards and development setbacks. Low human security means vulnerability to disasters. In his study of flood hazards, Roger Few (2003, 49) finds that "disruption to assets and livelihoods by one event often make households yet more vulnerable to future hazards." Citing Chambers (1995), Few continues, "The same families tend to lose their homes, possessions, livelihoods, increasing their vulnerability to the next disaster event."

The etymology of the word "disaster" indicates the traditional, astrological perception that disasters are beyond human control: from the Italian *dis* ("away, without") and Latin *astro* ("star, planet"). More recently, proactive efforts to reduce disasters have proven successful, as vulnerability has increasingly informed the understanding of disasters. Vulnerability is a function of factors (physical, social, economic, and environmental) that increase susceptibility to disaster. Such recognition vastly strengthens our understanding of disasters.

Recent efforts to reduce community vulnerability have appeared at all levels of civil society: the United Nations declared the 1990s the International Decade for Natural Disaster Reduction (IDNDR); in Gujarat, *panchayats* (village councils) formed networks and shared their time-tested practices in water harvesting for drought security. These efforts conceptualize vulnerability in different ways and thus attempt to reduce it through varied means. In "'Vulnerability': A Matter of Perception," Annelies Heijmans (2001) distinguishes three common views and "strategies" for addressing them (Table 3.1). It is apparent that the perspective of vulnerability directly influences the intervention strategies. For example, the social structural perspective prescribes political solutions.

Heijmans's strategies reduce vulnerability or further development. AIDMI has developed a victim security matrix (see Twigg 2001) that combines views on nature, poverty, and social structures and conceives of vulnerability in terms of security in four sectors: food, shelter, water, and livelihood. Nathan (2005) extends this holistic view and links it with five sectoral securities: livelihood and its resilience (assets and income-earning activities), base-line status or well-being (health and nutrition), self-protection (quality of house construction and location), social protection (by society), and governance (power system, rights, status of civil society).

Disasters set back community development. They may be initiated by natural or human-made hazards, but the underlying cause is human action. Development strategies attempting to reduce vulnerability focus on one or more aspect of human security.

Table 3.1
Views on vulnerability and resulting strategies for reduction

Cause	Solutions
Nature	Technological, scientific
Cost	Economic, financial
Societal structures	Political

Source: A. Heijmans and L. Victoria. 2001. "Citizenry-based and development oriented disaster management experiences and practices of the citizen's disaster response network in the Philippines." Manila, Philippines: Center for Disaster Preparedness: 118.

To sum up, "Reducing vulnerability of the poor is a development question," according to Fred Cuny (1983). He sees "disasters" as devastating events that may also serve as opportunities for communities to rise, organize, and cope. In this context, Razmara and Kaur (2000) point out that reducing vulnerability goes beyond policy to improve the living standard of the poor. It places a safety net that keeps the non-poor from falling into poverty and ameliorates the living standards of poor and non-poor alike (Razmara and Kaur, 2000). Cuny (1983, 86) explains that, "contrary to popular belief, a crisis reinforces local coping mechanisms and that local organisations often work better in times of crisis than in normal periods." The response to Hurricane Katrina in 2005 in the southern United States showed a modest level of interaction among institutions, but greater action at the local level (Louise and Thomas, 2006).

Against this backdrop, the general purpose of this chapter is to elucidate the significance of community-based strategies to reduce disaster risk in development and disaster management. It analyses the case of communities in rural Gujarat, India, specifically in the context of earthquake hazard.

Challenges in Providing Relief

To try to ameliorate destruction caused by natural calamities, families, communities, outside agencies, and governments provide humanitarian relief. Yet valuable emergency relief is difficult to provide from outside in ways that promote human security and reduce the likelihood of the disaster recurring. This section summarizes findings and indicates that, in the absence of community participation, disaster relief often inadvertently rebuilds the structures of vulnerability.

Two characteristics of outside humanitarian assistance may limit its effectiveness. First, emergency relief as a temporary injection to jump-start development has failed to reach its targets. In an assessment of relief in the flooding in Mozambique in 2000, Zefanias Matsimbe (2003, 35) recognizes that "emergency response and relief was largely designed to be a temporary solution to the crisis, and this actually impeded the sustainable recovery of at-risk communities." Haque and Burton (2005) and Middleton and O'Keefe (1998) explain that the entire top-down approach is inappropriate, as it aims to restore normality. Haque and Burton claim, "The restorations of order and so-called 'normal' conditions become the primary mission of crisis and disaster management, relief, and reconstruction" (352). Yet this very set of structures and processes resulted in the original disaster.

Second, some relief programs are top-down in structure. Mulwanda (1992, 89) observes that, despite the theoretical understanding of the advantages of participation, intervening agencies "continue to operate independently of victims resulting in the widening of the gulf between agencies and victims." Joachim and Patrick (2006) believe that, in the absence of a participatory approach to disaster management, the gap between response teams and sufferers continues to grow. Instead of empowering victims, relief occurs with little participation.

These two characteristics obscure the nature of relief. First, the idea of disasters as exceptional events, resulting from "an unfavourable position of a planet," means the needs of relief are often short-term and off-target. Second, relief initiatives often lack a solid community base.

Fred Cuny recognized in the early 1980s that "the growing awareness of the connection between disaster response and development is the single most important trend in disaster programs today" (1983, 257). In their study of organizational learning among nongovernmental organizations (NGOs) active in disaster mitigation, Twigg and Steiner (2002, 474) find "indications of shifts in attitude, with the old view of disasters as one-off events being replaced by awareness that development processes can influence the impact of disasters." This outcome suggests that extensive academic debates in recent decades (Blaikie et al. 1994; Burton 2005) have informed

NGOs' thinking. Interest has been growing in the nexus linking disaster risk reduction, climate change, and development (Sperling and Szekely 2005). Concerted efforts may be able to reduce the risk associated with climate change and extreme weather, which ultimately would lead to sustainable development (Schipper and Pelling 2006).

The following section discusses this progress in terms of community efforts to reduce disaster risk.

Progress: Understanding, Rhetoric, and Practice

As stakeholders gained a more dynamic understanding of disasters and vulnerability, they became increasingly able to target efforts appropriately for mitigation. Activities increased, especially vis-à-vis community self-reliance, management, development, and participation. Despite growing recognition of the need for such community participation, little institutionalization was taking place.

In the 1980s, Cuny (1983), Wijkman and Timberlake (1984), Maskrey (1989), and Anderson and Woodrow (1998) contributed to a more dynamic understanding of vulnerability. "The relationship between human actions and the effects of disasters – the socio-economic dimension of vulnerability – was increasingly well documented and argued, together with its implications for disaster management" (Twigg et al. 2000, 17). Vulnerability and capacity analysis, developed by Anderson and Woodrow, implies the need for relief agents to recognize the capabilities of the communities they intend to assist. This was one of several contributions in the period that added a developmental component to disaster relief.

Traditionally, people have generated strategies to cope with natural hazards. In Gujarat, India, communities harvest rainwater by saving it from the monsoon in catchments for use in dry seasons (AIDMI 2005). In Bangladesh, agricultural coping measures in flood-prone areas include selecting varieties of rice suitable for the season, soil, and moisture. The cultivation of banana and bamboo plants is widespread, as people can use them to construct rafts that keep livestock and assets afloat; riverbank erosion is controlled by planting *catkin* grass (Matin and Taher 2000). "People also tend to reduce the magnitude of economic loss in crops from floods and erosion, by cultivating low cost varieties [of rice] or late growing varieties" (ITDG n.d., 3). Rautela (2005) observes that farmers, to cope with dry spells, select crop varieties that resist long dry spells or the failure or delay of rain.

In 1994, the UNDP published the *Human Development Report*, and security issues have been recognized as the central theme. This document speaks of "human security," which emphasizes minimum or basic needs rather than either disaster response or development (Vaux 2005). Vaux further explains the relationship between poverty and security: the problem for the poorest people is not poverty, but insecurity; it is a matter not of becoming richer but of escaping destitution (Vaux 2005, 3).

In the humanitarian sector, the new concepts of vulnerability, security, capacities, and so on have affected the scope of activities by relief and development agencies. A few such organizations began including activities aimed at mitigation and preparedness and started to see certain activities as mitigation. Twigg et al. (2000) identify nine disaster mitigation and preparedness activities carried out by different NGOs: advocacy, environment, food security, housing, information, microfinance, preparedness, research, and training. Atlay and Green (2006) add eight items: building codes, emergency infrastructure protection and recovery of lifeline services, emergency rescue and medical care, evacuation of threatened populations, fatality management, insurance, land use zoning, and tax (dis)incentives. Many of these aim to support education (both for communities and for agencies) and self-reliance.

Instances of disaster preparedness have emerged in many sectors and across the world. "In Ghana, grain banks ... have made people less vulnerable to food shortages, eliminated exploitation of farmers by middlemen, and increased their market opportunities" (Yates et al. 2002, 2). "The owner of a sweetshop in Indore, India, interviewed in 1994, paid Rs 25 to put stepping stones around his shop so that customers would not have to stand in flood water. He said that if he had not done so it would have cost him Rs 100–200 in lost business" (Twigg 2002, 5). Similar efficiency has been seen on large scales: "The World Bank and US Geological Survey calculated that economic losses worldwide from natural disasters during the 1990s could be reduced by US$280 billion if US$40 billion were invested in mitigation and preparedness" (Twigg 2002, 5).

Despite increased global stress on disaster preparedness, practice has been less inspiring. Examples of community disaster risk reduction remain specific to project, area, disaster, or donor. In an investigation of 75 such projects, Twigg et al. (2000, 30) find that 54 were country-specific, most (61 per cent) addressed a single hazard, and "many, if not most," were "stimulated by recent disaster experience."

Unfortunately, "debates over alternative development models tended to overshadow the original calls to strengthen risk management" (Rocha and Christoplos 2001, 240). According to Wisner (2001, 251), "despite the rhetoric in favour of 'learning the lessons of Mitch', very little mitigation and prevention had actually been put in place between the hurricane (1998) and the earthquakes (2001)."

The IDNDR (the 1990s) was symbolic of the need for disaster mitigation. Yet, despite NGOs' comparative advantage in disaster mitigation and preparedness (DMP), which was widely recognized (e.g., Twigg et al. 2000; Matin and Taher 2001), the decade focused on governments and had little influence on NGOs. Twigg et al. (2000) show that 87 per cent of relief and development NGOs interviewed in the late 1990s about DMP had either not heard of the IDNDR or said that it had "no impact on work" (125). Thus at least one group of stakeholders with close ties to vulnerable communities was virtually left out of the international program.

Fortunately, this began to change with the Yokohama Conference in 1994, which emphasized community programs to reduce vulnerability and NGOs managing natural hazards (Twigg 2001, 126). Pelling (2003) argues that community approaches reveal the local dynamics and intricacies of vulnerability, besides addressing vulnerability and strengthening local capacities. Khan (2004) attributes most effective and successful disaster reduction to community participation and involvement. However, these efforts require external support from NGOs.

The increased commitment to local organizations at Yokohama symbolized the broad recognition of prevention and revealed stakeholders at various levels acting simultaneously. It became clear that each stakeholder could help and popularize risk reduction (La Trobe and Davis 2005). Only new policies can change systems that have led to disenfranchisement and vulnerability (Middleton and O'Keefe 1998). In this context, Garber (2007) notes that the shortfall in the US federal response to Katrina was the fault of policy and political constraints, which impeded disaster management. Roles are decentralized in the United States roughly according to comparative advantage. The advantages include "local perspectives into policy making and rural planning, two-way communication with higher policy levels, implementation of rural development activities at local level, mobilizing local participation, and handling emergencies at the local level with conscious links to reconstruction, prevention and preparedness phases of disaster management" (ADPC 2003, ii).

The role of communities now appears central to sustainable development in both academic literature (Burkey 1993; Manojlovic and Pasche 2008; Lawther 2009) and practical agency publications. The concept of community disaster management has gained widespread recognition – for example, in Bangladesh and in China. The emphasis has been on use of local expertise and knowledge to select suitable responses (Gurr and Harff 2003; Thomalla et al. 2006). Local and international organizations alike have discussed the application of community-based "strategies" or processes for disaster prevention: "Those who have undertaken participatory work have learned that local communities have capabilities that outsiders are largely ignorant of – capabilities for appraisal and analysis of their circumstances, for planning measures to meet their needs, for implementing and managing projects, and for evaluating the results. Moreover, information gathered and shared through participatory approaches has proved to be valid and reliable" (Twigg et al. 2001, 6).

The broad recognition of the relationship between community participation and sustainability has led more organizations to support local processes in disaster prevention. Sometimes, however, organizations design projects to meet their own needs and then try to stimulate community involvement (IFRC 2001 in Christoplos et al. 2001). Christoplos et al. explain (2001: 192) that such putting the "cart-before-the-horse ... is common and exacerbates opportunities for elite capture of resources intended for the most vulnerable." The scope for improving outsiders' understanding of community-based disaster risk reduction (CBDRR) is apparent. Details of CBDRR and its opportunities appear in the next section below.

Difficulties in supporting community approaches are considerable. Many organizations involved in disaster management maintain a top-down approach because of the urgency of disasters and responses as well as entrenched systems of power and the division of resources. Twigg et al. (2001, 5) explain that people in positions of power, whether political, institutional, or professional, are "reluctant to hand over authority to the grass roots. Many organizations have called their work 'participatory' but have not changed their approach." He goes on: "Participation stands in contrast to more traditional practice where external actors collect data, define problems, design projects and assess their results. This 'topdown' approach has been heavily, and increasingly, criticised since the late 1970s, on the grounds that it tends not to reach those worst affected by disaster, may even make them more vulnerable, can be manipulated by political interests, is often inefficient, usually takes a unisectoral approach and does not respond to people's real needs" (7).

Additionally, participatory approaches often require a lot of time and resources. The most visible actors – the public and non-profit sectors – must redirect resources from development programs. "It is hard to gain votes by pointing out that disasters *did not* happen" (Christoplos et al. 2001, 195). Rocha and Christoplos (2001) cite post-Mitch Nicaragua as an example of how emergencies encourage local and national governments to reform policies that made communities vulnerable. The structure of most funding is another disincentive for communities and organizations to engage in prevention. Non-market-driven organizations, at lower levels, often rely on donors to support prevention. The bulk of this funding becomes available in the wake of large disasters.

The following section goes deeper into CBDRR. Citing development literature and the practical experience of organizations dealing with disasters, it shows that CBDRR provides opportunities to overcome problems caused by typical top-down, supply-driven, post-disaster responses.

DISASTER MITIGATION: EFFECTIVE
WHEN COMMUNITY-BASED

What Is "Community-Based" Disaster Risk Reduction?

The United Nations International Strategy for Disaster Reduction (UNISDR) (2002, 17) defines disaster risk reduction as "the conceptual framework of elements considered with the possibilities to minimize vulnerabilities and disaster risks throughout a society, to avoid (prevention) or to limit (mitigation and preparedness) the adverse impacts of hazards, within the broad context of sustainable development."

As we saw above, sustainable risk reduction must engage the primary beneficiaries. The International Institute for Disaster Risk Management describes community-based disaster management as "an approach that involves direct participation of the people most likely to be exposed to hazards, in planning, decision making, and operational activities at all levels of disaster management responsibility" (UNCRD 2003, 11). This definition indicates the primacy of "people most likely to be exposed to hazards" but does not specify the requisite extent of participation. Instead of employing "an approach that involves direct participation," "community-based" methods depend on community control. Correspondingly, we propose another working definition: the process through which people most vulnerable to hazards exercise control of, or primary influence in, initiatives that decrease their susceptibility to disasters.

Lorna Victoria (2002, 276), who provides an "orientation on the why, what, who, when, how and so what" of CBDRR, identifies seven basic elements:

- people's participation
- priority for the most vulnerable groups, families, and people in the community
- community-specific risk reduction
- recognition of existing capacities and coping mechanisms
- reduction of vulnerabilities by strengthening capacities so as to build a disaster-resilient community
- linking disaster risk reduction with development
- supporting and facilitating roles for outsiders

Such a characterization helps to distinguish activities that are genuinely community-based from those simply given the name. This also helps practitioners identify opportunities for supporting them.

How Do Community-Based Strategies Reduce Risk?

Strategies for risk reduction are most effective when based on the realities of the group they seek to serve. This is a function of six reasons: appropriate and sustainable efforts, indigenous knowledge and capabilities, cost-effectiveness/replication, response/first responders, democratic participation, and appropriate involvement for others:

i *Appropriate and sustainable efforts*: "Because they draw deeply from the community's perception of its vulnerability, past disaster experiences, awareness of their own resources, and connection with other development priorities[,] these grassroots efforts are extremely valuable and most likely to be sustained over longer periods" (Moga 2002, 172). Further, "coping strategies based on household resource management and adaptability have been far more significant than any external assistance" (Matin and Taher 2001, 237). Indigenous and community-based (CB) networks are often "deemed more suitable for initiating the first response" because they have good background knowledge of risk-prone areas and are able to identify socially vulnerable groups (Matin and Taher 2000, 4). "Involvement of local people promotes self-reliance and ensures that emergency management plans meet local needs and circumstances" (Pusch 2004, 24). Community-based disaster risk reduction (CBDRR) helps maximize the potential utility of local contributions, resources, and capacities.

ii *Indigenous knowledge and capabilities*: People in disaster-prone areas have developed resilience to cope with disasters through indigenous knowledge and means. Based on their extensive experience, people in disaster-affected environments have learned to cope by using their indigenous techniques (Matin and Taher 2001). Community "insiders" therefore can inform "outsiders" of hazard risks and mobilize the community or groups to reduce their vulnerability (OECD 1994). Some observers believe outsiders are the ignorant actors (in relief and development work) and local community members the experts (Chambers 1997 in Twigg et al. 2001, 11).

iii *Cost-effectiveness/replication*: Successful local initiatives at disaster mitigation are likely to spread to other communities. The costs of spreading such measures fall to users. Zenaida Delica-Willison, of the Centre for Disaster Preparedness (Philippines), considers community risk management "the single most important tool available for us today in reducing the increasing cost related to natural disaster" (UNCRD 2004, 84).

iv *Response/first (sometimes only) responders*: As families, neighbours, and other locals are nearby, they are often the first, and sometimes only, responders. "The effects of a disaster are first felt at the level of the community, and the community is the first to respond to a disaster. The greatest numbers of lives are saved during the first few hours after a disaster occurs, before outsiders arrive" (Pusch 2004, 24). Since disasters affect livelihoods, the involvement of individuals and the community can reduce the impact (UNCRD 2004, 5).

v *Democratic participation*: "Preparation at the community level improves a nation's capabilities in responding to and coping with disasters, as local communities form the building blocks of a country. Involving communities creates a pressure group to ensure that authorities at both the local and central levels remain responsive to community concerns. A community-level focus facilitates identification of vulnerable groups, such as women, the elderly and ethnic groups. The concerns of these groups should be voiced in any participatory effort" (Pusch 2004, 25).

vi *Appropriate involvement for others*: With advocacy and local experience, local organizations may provide feedback relevant for establishing human security in the policy loop. The community approach can help avoid some recognized disadvantages of external disaster relief, such as the disincentive to self-reliance and disruption to local produce markets resulting from distribution of free food. CBDRR places greater stress on "what can be learned" and how best to "enable, sustain and scale it up" (UNCRD 2004, 84). "In some cases, participation is an end in itself, enabling men and women to learn, organise, decide, plan and take action without other specific goals in mind" (Twigg et al. 2000, 25).

Challenges to a Community-Based Approach

Community initiatives to reduce disaster risk aim to enable/support communities in activities that will protect them against hazards. Authentic participation is a means to this end. However, several aspects of community initiatives, both within and outside the community, may inhibit their recognition and wide practice.

External inhibitors are isolation, lack of national recognition (UNCRD 2003), and tension between natural community processes and project design by "support" agencies. Because CBDRR occurs at the ground level, lessons from one community may not work for others or may need adaptation. Additionally, senior-level underestimation of local initiatives may inhibit the development of supportive infrastructure and policies.

Fowler (1997) notes the inappropriateness of agency mandates and timetables for more spontaneous community processes. Additionally, Twigg et al. (2001) write about agency personnel, "Disaster specialists have been slower to take to participatory approaches than their colleagues in development. This is largely due to the history, character and culture of disaster work, with its command-and-control mentality, blueprint planning, technocratic bias and disregard for vulnerable communities' knowledge and expertise regarding hazards and methods of disaster mitigation. We can see one dimension of this in the relationship between official organisations and 'emergent' groups after a disaster ... emergent groups are valuable as well as ubiquitous, yet they may bother disaster managers because they are outside their plans, systems and – above all – control" (11).

Challenges also come from within the community. Sharp disagreements among groups may cause conflicts of interest regarding whose risk matters, where mitigating infrastructure belongs, and who should lead the planning (Burkey 1993). Twigg et al. (2001) discuss similar challenges and how outsiders may facilitate participation.

Numerous organizations have been involved with disaster preparedness; however, fewer have focused on CBDRR. Some of the prominent examples are the Asian Disaster Preparedness Centre (ADPC) (www.adpc.net), the All India Disaster Mitigation Institute (AIDMI) (www.southasiadisasters.net), Duryog Nivaran (www.duryognivaran.org), the International Federation of Red Cross and Red Crescent Societies (IFRC) (www.ifrc.org), and the International Strategy for Disaster Reduction (ISDR) (www.unisdr.org). Others include South Africa's Peri Peri, LA RED, and the Institute for Disaster Risk Management (which has an accredited course on Community Level Disaster Risk Management). Some NGOs, including ACTIONAID, CARE, CONCERN, and OXFAM, have their own versions of CBDRR in different countries (UNCRD 2004, 13).

CONCLUSION AND RECOMMENDATIONS

What We Know

In recognizing the effectiveness and sustainability of local initiatives to reduce disaster risk, efforts in different regions have increasingly focused on mitigation and preparedness. The devil, however, lies in the details. Despite wide recognition of the necessity of community "ownership" and beneficiaries' participation, efforts to support genuine community involvement are rare. Existing endeavours are often specific to a project, area, disaster, or donor. They are rarely institutionalized or sustainable.

Twigg and Steiner (2002, 474) explain that consideration of vulnerability and mitigation is beginning to "penetrate NGO consciousness at the policy level but this is not translated to the operational level, where disaster risk reduction activity tends to be sporadic, poorly integrated with development planning, and largely unsupported by institutional structures and systems."

Some community institutions in the South have been reducing disaster risk for years. Their efforts include researching and documenting the realities of these local initiatives. Organizations such as AIDMI learn from their own operations with disaster-prone communities and make the lessons available for support elsewhere. Their endeavours confirm that disaster mitigation can succeed when it involves people who have most to gain.

What We Can Do

The discussion above centres on the importance of a community base for reducing disaster risk among the vulnerable. If local groups are to lead their own efforts, what role remains for those who recognize the importance of reducing risk everywhere? In considering participation, self-reliance, and community development, partnership with community members and local institutions is essential (Burkey 1993; Fowler 1997; Twigg 2004). The organizations discussed above are conducting programs directly with communities and supporting CBDRR. Below, we summarize two types of activities – building institutions and capacities and managing knowledge – especially appropriate for stakeholders in partnerships with local communities. Topics overlap because the categories are interrelated.

Institution/Capacity-Building

SUPPORTING LOCAL ORGANIZATIONS. Local organizations working with communities have a more nuanced understanding of the context than outsiders do. They can provide support services in local languages and in recognition of local customs. These organizations have staff with expertise and ability to create tools for, elicit insights from, and engage in activities with the communities they serve. They can help outsiders conduct assessments and design and implement mitigation programs in line with community needs. Support can be given to local institutions in terms of strategy development, resources, and staff training. North-South and South-South partnerships between organizations with similar objectives offer opportunities for exchanging information, staff, and ideas.

SUPPORTING COPING MECHANISMS. People across the world have designed personalized manners of reducing the effects of micro-disasters in their lives. Such mechanisms deserve recognition and range from strength in family relations to intermediate technologies and stockpile systems. Establishments such as AIDMI's chamber of commerce and industry for small businesses have grown around people's inspiration to establish human security. "[Efforts] designed to build up the assets of the poor to withstand shocks will be increasingly important in reducing the human burden from [disasters]" (Sanderson 2000 in Few 2003). Strengthening coping mechanisms is recognized as an appropriate method for reducing vulnerability. This can be done by identifying mechanisms effective in local risk reduction; demonstrating their effectiveness in areas with similar hazards; supporting communities in designing a measure; or providing support to coping strategies demonstrated during disasters. Additionally, outside organizations may support individuals in their ability to make decisions and manage community activities. Partners may also help develop local evaluation and management skills for emergency response. Relations with a local organization will greatly increase the efficiency of each of these.

SUPPORTING LIVELIHOODS. The understanding that strong livelihoods are integral to disaster resilience has been established. The recent emergence of microfinance and cash-for-work schemes for risk reduction is expected to increase local demand for support service. Identifying needs and the potential application of these schemes and similar ones will continue to be important.

Knowledge Management

PROVIDING TRAINING. When people have information from a variety of sources, they are better able to plan for themselves, make informed choices, and act to reduce their own vulnerability (IFRC 1995 in Pearce 2003). Too few organizations provide training related to bottom-up risk reduction. Even fewer do so in local languages, which would achieve the strongest impact on vulnerable communities. Training programs, such as the ADPC and AIDMI's, have been designed with continuous input from participants. Similar activities should be replicated.

ADVOCACY. Policy initiatives require input from local experiences. Lessons and ideas from the disaster-affected and vulnerable population should be recorded and channelled to decision-makers. Similarly, the vulnerable should be supported in influencing policy decisions that affect their lives. When accumulated and used appropriately, this knowledge and the people whose experiences it represents may serve as leverage on institutions that can promulgate legislation.

REFERENCES

ADPC (Asian Disaster Preparednes Center). 2004. *Program Completion Report: Asian Urban Disaster Mitigation Program*. Bangkok: ADPC.

All India Disaster Mitigation Institute (AIDMI). 2005. *Moving towards Water Security: Structures, Tools, and Processes for Drought-Prone Areas*. Ahmedabad: AIDMI with EU.

Altay, N., and W.G. Green. 2006. "Interfaces with other disciplines: OR/MS research in disaster operations management." *European Journal of Operational Research* 175(1): 475–93.

Anderson, M., and P. Woodrow. 1998. *Rising from the Ashes: Development Strategies in Times of Disaster*. London: IT Publications.

Blaikie, P., T. Cannon, I. Davis, and B. Wisner. 1994. *At Risk: Natural Hazards, People's Vulnerability, and Disasters*. New York/London: Routledge.

Burkey, S. 1993. *People First: A Guide to Self-Reliant, Participatory Development*. London: Zed Books.

Burton, I. 2005. "The social construction of natural disasters: an evolutionary perspective." In *Know Risk*. Geneva: United Nations: 35–6.

Chambers, R. 1995. "Poverty and livelihoods: whose reality counts?" *Environment and Urbanization* 7 (1): 173–204.

– 1997. *Whose Reality Counts? Putting the First Last*. London: Intermediate Technology Publications.

Christoplos, I., J. Mitchell, and A. Liljelund. 2001. "Re-framing risk: the changing context of disaster mitigation and preparedness." *Disasters* 25(3): 185–98.

Cuny, F. 1983. *Disasters and Development*. New York/London: Oxford University Press.

Few, R. 2003. "Flooding, vulnerability and coping strategies: local responses to a global threat." *Progress in Development Studies* 3 (1): 43–58.

Fowler, A. 1997. *Striking a Balance*. London: Earthscan.

Garber, B.J. 2007. "Disaster management in the United States: examining key political and policy challenges." *Policy Studies Journal* 35 (2): 227–38.

Gurr, T.R., and B. Harff. 2003. "From surveillance to risk assessment to action." In K.M. Cahill, ed., *Emergency Relief Operations*. New York: Fordham University Press: 3–31.

Haque, C.E., and I. Burton. 2005. "Adaptation options strategies for hazards and vulnerability mitigation: an international perspective." *Mitigation and Adaptation Strategies for Global Change* 10 (3): 335–53.

Heijmans. A., and L. Victoria. 2001. "Citizenry-based and development oriented disaster management experiences and practices of the citizen's disaster response network in the Philippines." Manila, Philippines: Center for Disaster Preparedness: 118.

IFRC (International Federation of Red Cross). 2001. *World Disasters Report 2001: Focus on Recovery*. Oxford: Oxford University Press.

Joachim, A., and M.R. Patrick. 2006. "The importance of governance in risk reduction and disaster management." *Journal of Contingencies and Crisis Management* 14 (4): 207–20.

Khan, S. 2004. "Vietnam case study." Paper presented at the international symposium Community Legacy in Disaster Management, Kobe, 7 Feb.

La Trobe, S., and I. Davis. 2005. *Mainstreaming Disaster Risk Reduction: A Tool for Development Organisations*. Middlesex: Tearfund. www.proventionconsortium.org/files/Tearfund_mainstreaming_DRR.pdf

Lawther, P.M. 2009. "Community involvement in post disaster reconstruction: case study of the British Red Cross Maldives recovery program." *International Journal of Strategic Property Management* 13 (2): 135–69.

Louise, K.C., and W.H. Thomas. 2006. "Communication, coherence, and collective action: the impact of Hurricane Katrina on communications infrastructure." *Public Works Management and Policy* 10 (4): 328–43.

Manojlovic, N., and E. Pasche. 2008. "Integration of resiliency measures into flood risk management concepts of communities." *WIT Transctions on Ecology and the Environment* 118: 235–45.

Maskrey, A. 1989. *Disaster Mitigation: A Community-Based Approach*. Oxford: Oxfam.

Matin, N., and M. Taher. 2000. "Disaster mitigation in Bangladesh: country case study of NGO activities." *Report for Research Project "NGO Natural Disaster Mitigation and Preparedness Projects: An Assessment and Way Forward."* ESCOR Award No. R7231. London: British Red Cross.

– 2001. "The changing emphasis of disasters in Bangladesh NGOs." *Disasters* 25 (3): 227–39.

Matsimbe, Z. 2003. *Assessing the Role of Local Institutions in Reducing the Vulnerability of At-Risk Communities in Búzi, Central Mozambique*. Cape Town: DiMP, University of Cape Town.

Middleton, N., and P. O'Keefe. 1998. *Disaster and Development*. London/ Chicago: Pluto Press.

Moga, J. 2002. "Disaster mitigation planning: the growth of local partnerships for disaster reduction." Paper for Regional Workshop on Best Practices in Disaster Mitigation. Bangkok: ADPC.

Mulwanda, M. 1992. "Active participants or passive observers?" *Urban Studies* 29 (1): 89–97.

Nathan, F. 2005. "Vulnerabilities to natural hazard: case study on landslide risks in La Paz." Paper for the World International Studies Conference (WISC) at Bilgi University, Istanbul, Turkey, 24–27 Aug.

Organisation for Economic Cooperation and Development (OECD). 1994. *Guidelines on Aid and Environment No. 7: Guidelines for Aid Agencies on Disaster Mitigation*. Paris: OECD.

Pearce, L. 2003. "Disaster management and community planning, and public participation: how to achieve sustainable hazard mitigation." *Natural Hazards* 28: 211–28.

Pelling, M. 2003. *Natural Disasters and Development in a Globalising World*. London: Routledge.

Pusch, C. 2004. "Preventable losses: saving lives and property through hazard risk management: a comprehensive risk management framework for Europe and Central Asia." World Bank Disaster Risk Management Working Paper Series No. 9. Washington, DC: World Bank.

Rautela, P. 2005. "Indigenous technical knowledge inputs for effective disaster management in the fragile Himalayan ecosystem." *Disaster Prevention and Management* 14 (2): 233–41.

Razmara, S., and I. Kaur. 2000. *Reducing Vulnerability and Increasing Opportunity: A Strategy for Social Protection in Middle East and North Africa*. Report for Social Protection Group of Middle East and Northern Africa Region. Washington, DC.

Rocha, J., and I. Christoplos. 2001. "Disaster mitigation and preparedness on the post-Mitch agenda." *Disasters* 25 (3): 240–50.

Schipper, L. and M. Pelling. 2006. "Disaster risk, climate change and international development: scope for, and challenges to, integration." *Disasters* 30 (1): 19–38.

Sperling, F., and F. Szekely. 2005. "Disaster risk management in a changing climate." Informal discussion paper for the World Conference on Disaster Reduction on behalf of the Vulnerability and Adaptation Resource Group (VARG), Washington, DC.

Thomalla, F., T. Downing, E. Spanger-Siegfried, G. Han, and J. Rockstrom. 2006. "Reducing hazard vulnerability: towards a common approach between disaster risk reduction and climate adaptation." *Disasters* 30 (1): 39–48.

Twigg, J. 2001. "Sustainable livelihoods and vulnerability to disasters." Disaster Management Working Paper 2/2001. Benfield Greig Hazard Research Centre, University College London. www.bghrc.com

– 2002. "Lessons from disaster preparedness." Notes for presentation to International Conference on Climate Change and Disaster Preparedness, The Hague, Netherlands, 26–28 June, Workshop 3: It Pays to Prepare. Benfield Greig Hazard Research Centre, University College London. www.bghrc.com

– 2004. *Disaster Risk Reduction: Mitigation and Preparedness in Development and Emergency Programming*. London: Humanitarian Policy Group/ODI.

– M. Bhatt, A. Eyre, R. Jones, E. Luna, K. Murwira, J. Sato, and B. Wisner. 2001. "Guidance notes on participation and accountability (draft)." Benfield Greig Hazard Research Centre, University College London. www.bghrc.com

Twigg, J., and D. Steiner. 2002. "Mainstreaming disaster mitigation: challenges to organisational learning in NGOs." *Development in Practice* 12 (2 and 3): 473–9.

Twigg, J., D. Steiner, M. Myers, and C. Benson. 2000. "NGO natural disaster mitigation and preparedness projects." Report for Research Project "NGO Natural Disaster Mitigation and Preparedness Projects: An Assessment and Way Forward." ESCOR Award No. R7231. London: British Red Cross.

United Nations Centre for Regional Development (UNCRD). 2003. *Sustainability in Grass-Roots Initiatives: Focus on Community-Based Disaster Management*. Hyogo, Japan: UNCRD.

– 2004a. *Sustainable Community Based Disaster Management Practices in Asia: A Users Guide*. Hyogo: UNCRD.

– 2004b. *Tapestry*. Hyogo: UNCRD.

UNDP (United Nations Development Programme). 2005. "Planning development today for a world with fewer disasters tomorrow." In *Know Risk*. Geneva: United Nations: 33–4.

– 1994. *Human Development Report 1994*. New York: UNDP.

United Nations International Strategy for Disaster Reduction (UNISDR). 2002. *Living with Risk: A Global Review of Disaster Reduction Initiatives*. Geneva: UNISDR.

Vaux, T. 2005. *Beyond Relief*. Ahmedabad: AIDMI.

Victoria, L. 2002. "Community-based approaches to disaster mitigation." Paper for Regional Workshop on Best Practices in Disaster Mitigation. Bangkok: ADPC.

Wijkman, A., and L. Timberlake. 1984. *Natural Disasters: Acts of God or Acts of Man?* London: Earthscan.

Wisner, B. 2001. "Risk and the neoliberal state: why post-Mitch lessons didn't reduce El Salvador's earthquake losses." *Disasters* 25 (3): 251–68.

Yates, R., K. Alam, J. Twigg, D. Guha-Sapir, and P. Hoyois. 2002. "Development at risk." Brief for the World Summit on Sustainable Development, Johannesburg, South Africa, 26 Aug–4 Sept.

PART TWO

International Perspectives on Disaster Risk Management and Public Policies

4

The Intersection of Policies on Disaster Management, Climate Change, and International Development

G.A. MCBEAN

INTRODUCTION

Often the major news in international media is about how a hazardous event has devastated a country and set back development there. Although the occurrence may be an earthquake, most often it is a typhoon or hurricane, a flood or a drought, or some other catastrophe relating to weather or climate. The intersection of disasters, often relating to climate and its variability and change, on countries trying to further develop is the topic of this chapter.

At least since the 1990s, the annual number of natural disasters has been rising (Gall, Borden, and Cutter 2009; Munich Re 2010a; Vos et al. 2010). In 2010, about 260,000 people died in natural disasters (Borenstein and Reed Bel 2010); we can compare this to fewer than 115,000 deaths from terrorism in the period 1968–2009. A category-5 disaster is one with more than 500 deaths and/or overall losses of more than US$500 million. Whereas there were between 5 and 15 category-5 events per year in the 1980s, there were 15 to 25 for 1990–2005 and 28 to 41 in 2006–8 (Munich Re 2008). During 2000–8, more than 220 million people were victims of the roughly 360 climate-related disasters per year (Rodriguez et al. 2009). Floods and storms were the dominant factor in these disasters. Combinations of increases in population, poverty, and valuable and vulnerable infrastructure and a changing climate have led to these rises in disaster events. Wahlström (2009, 5) states: "Over the last two decades (1988–2007), 76% of all disaster events were hydrological, meteorological or climatological in nature; these accounted for 45% of the deaths and 79% of the economic losses caused by natural hazards."

The Intergovernmental Panel on Climate Change's (IPCC 2007b) full scientific assessment concludes: "Warming of the climate system is unequivocal" and "Most of the observed increase in global average temperatures since the mid-20th century is very likely due to the observed increase in anthropogenic greenhouse gas concentrations." Global mean temperatures over the past 25 years have been increasing at 0.18°C per decade, and there are no indications of a slowdown or pause in human-caused climatic warming (Copenhagen Diagnosis 2009). According to a report by the US National Academy of Sciences (Matson et al. 2010) (summarized by National Research Council 2010): "A strong, credible body of scientific evidence shows that climate change is occurring, is caused largely by human activities, and poses significant risks for a broad range of human and natural systems." Both scientists and insurers expect that, as the climate changes, there will be more frequent and intense extreme weather events, resulting in more costly disasters (Munich Re 2010b).

In December 2009, many national leaders (including Prime Minister Stephen Harper of Canada), ministers, and other officials in Copenhagen at the Fifteenth Conference of the Parties to the United Nations Framework Convention on Climate Change agreed to the Copenhagen Accord, with the opening paragraph: "We underline [the fact] that climate change is one of the greatest challenges of our time … We recognize the critical impacts of climate change and the potential impacts of response measures on countries particularly vulnerable to its adverse effects and stress the need to establish a comprehensive adaptation programme including international support."[1] The Cancun Agreement[2] of the follow-up gathering in December 2010 adopted similar wording.

Around the world, poverty, hunger, disease, illiteracy, environmental degradation, and discrimination against women continue to be major concerns. Within the broad range of commitments to human rights, good governance, and democracy, the world's leaders, at the Millennium Summit in 2000, passed a declaration that established a series of Millennium Development Goals (MDGs). The MDGs provided a comprehensive and multi-dimensional development framework and set clear, quantifiable targets to achieve in all countries by 2015 (Millennium Declaration and Goals 2008).

These three issues – disasters, climate change, and international development – are strongly linked, and the approaches of governments to them need to be equally strongly linked. This is the topic of this chapter.

POLICY FRAMEWORK

One of the most basic roles of government has long been protecting people through services such as national defence, law enforcement, and fire protection. Since "providing security for the nation and for its citizens remains the most important responsibility of government" (United Kingdom's National Security Strategy, 2008), defending citizens from the impact of natural disasters and a changing climate and responding to these would seem basic roles of government. Governments, however, must always balance a broad range of policy issues and make difficult decisions regarding how to allocate scarce public funds among the economy, protection, and development. These additional responsibilities should be seen in this context.

In the Speech from the Throne[3] on 3 March 2010, the Canadian government enunciated its positions on many issues. "Our Government will use its voice to speak on behalf of Canada's commitment to global security and human rights ... Nowhere is a commitment to principled policy, backed by action, needed more than in addressing climate change ... The Copenhagen Accord reflects these principles and is fully supported by the Government of Canada."

The management of natural disasters and the links with climate change and international development are public policy issues, and there should be at least links between them, if not an integrated policy framework. The concept of sustainable development, which has been adopted by many governments, can provide a framework for this integration. Though relating specifically to Canada, the conclusions of this chapter have much broader applicability.

DISASTER MANAGEMENT AND HAZARD MITIGATION

In 2005, governments attending the World Conference on Disaster Reduction (Kobe, Hyogo, Japan) declared: "We are deeply concerned that communities continue to experience excessive losses of precious human lives and valuable property as well as serious injuries and major displacements due to various disasters worldwide" and "We can and must further build the resilience of nations and communities to disasters" (Hyogo Declaration 2005[4]). The Hyogo Framework for Action (HFA) specified that governments must "ensure that disaster risk reduction is a national and a local priority with a strong institutional basis for implementation." In June 2009, as part of its commitment to deliver on the HFA, Canada announced the establishment of a National Platform for Disaster Risk Reduction to build multi-stakeholder coordinated leadership in reducing disaster risk.[5]

A hazard is a "potentially damaging physical event, phenomenon or human activity that may cause the loss of life or injury, property damage, social and economic disruption or environmental degradation."[6] A hazard triggers a disaster when it interacts with various forms of vulnerability (i.e., physical, social, economic, and environmental factors that increase the susceptibility of a community to hazard impacts). Disaster management refers to policies and practices developed and implemented to manage the impact of disasters. Effective disaster management requires extensive planning before a disaster, targeting four elements: preparedness (policies and procedures to facilitate effective response); response (actions immediately before, during, and after a disaster to protect people and property and to enhance recovery); recovery (actions after a disaster to restore critical systems and return a community to pre-disaster conditions); and mitigation (actions before or after a disaster to reduce the impact on people and property) (Godschalk 1991).

In the context of natural disasters, mitigation includes policies and actions to prevent or reduce losses associated with natural hazards. Means could include building codes and standards to protect people, property, and infrastructure from "reasonable" extremes; land use zoning to prevent inappropriate development in hazardous areas; structural engineering to increase resistance to hazard pressures (e.g., flood protection; reinforced concrete in earthquake zones); and forecasting and warning systems that provide information to people and advice about appropriate response (Godschalk et al. 1999).

Although all countries have been hit by natural disasters, usually more people die in developing countries and costs are greater in developed countries (Mileti 1999; Mizra 2003). The average number of deaths per disaster is 23 in highly developed countries, about 150 in the medium-range, and over 1,000 in the developing (Mutter 2005). While the costs in the most developed countries are large, they are much greater in the developing countries as a percentage of gross domestic product (GDP) (Handmer 2003). Among the key findings and recommendations of the UN ISDR's Global Assessment Report on Disaster Risk Reduction (2009) are:

- Global disaster risk is highly concentrated in poorer countries with weaker governance.
- Weather-related disaster risk is expanding rapidly in terms of territories affected, losses reported, and frequency of events.
- Climate change is already altering the geographical distribution, frequency, and intensity of weather-related hazards and threatens to undermine the ability of poorer countries and their citizens to absorb and recover from disaster.
- The governance arrangements for reducing disaster risk in many countries do not integrate risk considerations in development.

Within the government of Canada, coordinating emergency and disaster management falls to Public Safety Canada (PSC), a ministry created largely in response to the threats of international terrorism. After many years of consultation and deliberation with interested parties, the federal government has asserted its commitment to a National Disaster Mitigation Strategy, with the goal of "protecting lives and maintaining resilient, sustainable communities for fostering disaster risk reduction as a way of life" (Cullen 2004), and in January 2008 the federal, provincial, and territorial governments launched a National Disaster Mitigation Strategy.[7]

However, although hazard researchers generally agree that investment in mitigation saves far more than the cost to recover from disaster impacts (Mileti 1999; Canton 2007), most government expenditures in Canada target recovery, for various reasons (Henstra and McBean 2005). First, public demand for mitigation is generally weak. Many people continue to view natural disasters as "acts of God" and thus conclude that there is little that an individual can do to prevent losses. In other cases, people may be uninterested in loss prevention because they believe they are covered by insurance. From a political perspective, hazard mitigation is less appealing than other public investments: while the costs of mitigation are immediate and often substantial, the benefits are realized only in the case of another disaster, which may never occur. In a political system predicated on a roughly four-year election cycle, investment costs borne by one government (and set of taxpayers) may be realized only by a future government and a somewhat-different group of taxpayers.

Local governments can play an instrumental role in hazard mitigation, because they are close to hazards and control many of the most effective tools (e.g., regulating land use and enforcing building codes) (Prater and Lindell 2000). However, disasters in any particular community are statistically rare, local demand for disaster mitigation is minimal, and most of the burden for disaster recovery is shouldered by federal and provincial governments, so local governments have little economic or political incentive to invest in mitigation (Berke et al. 2008). Moreover, most post-disaster investments in mitigation would not be eligible for intergovernmental cost-sharing under Canada's current disaster recovery assistance regime. It is increasingly important that governments at all levels acknowledge the impact of natural disasters and implement better policies and programs for their management.

CLIMATE CHANGE

The objective of the UN Framework Convention on Climate Change,[8] which has been signed and ratified by almost every country in the world, is "stabilization of greenhouse gas concentrations in the atmosphere at a level that would prevent dangerous anthropogenic [human-induced] interference with the climate system. Such a level should be achieved within a time frame sufficient to allow ecosystems to adapt naturally to climate change, to ensure that food production is not threatened, and to enable economic development to proceed in a sustainable manner." Despite clear economic dimensions, the principles of the convention also include a substantial security element, which obligates governments to protect citizens. While part of the increase in natural disaster losses referred to above can be attributed to factors such as population growth, greater hazard exposure, and higher property value, changing weather conditions also play a role and are likely to become of greater significance. Observed changes in the frequency and severity of natural hazard events are of particular concern to insurers, and at the World Conference on Disaster Reduction in 2005 in Kobe, Hyogo, Japan, Munich Reinsurance Group proposed radical political, social, and economic measures to stem the escalating losses (Munich Re Group 2005).

Table 4.1
Recent trends and projections of extreme weather and climate events for which there is evidence of an observed late-twentieth-century trend

Phenomenon and direction of trend	Likelihood that trend occurred in late 20th century (typically since 1960)	Likelihood of future trend based on projections for 21st century using SRES scenarios
Warm spells/heat waves: frequency increases over most land areas.	Likely	Very likely
Heavy precipitation: frequency (or proportion of total rainfall from heavy falls) increases over most areas.	Likely	Very likely
Area affected by droughts increases.	Likely in many regions since 1970s	Likely
Intense tropical cyclone activity increases.	Likely in some regions since 1970	Likely
Incidence of extreme high sea level (excludes tsunamis) increases.	Likely	Likely

Note: The IPCC used the following terms to indicate assessed likelihood, using expert judgment, of an outcome or a result: Very likely > 90%, Likely > 66%.

Source: Table TS.4 from S. Solomon, D. Qin, M. Manning, et al. 2007. "Technical summary." In S. Solomon, D. Qin, M. Manning, Z. Chen, M. Marquis, K.B. Averyt, M. Tignor, and H.L. Miller, eds., *Climate Change 2007: The Physical Science Basis. Contribution of Working Group I to the Fourth Assessment Report of the Intergovernmental Panel on Climate Change.* Cambridge and New York: Cambridge University Press: Table TS4: 52.

Climate change is expected to alter our physical environment in a number of ways, including more frequent and intense hazard events, rising sea levels, and various effects of changing temperature. The IPCC has noted that human systems are vulnerable to climate extremes, as demonstrated by many natural disasters and the likelihood that, despite uncertainty about future frequencies of extreme events, the impact will probably increase with climate change (Solomon et al. 2007). Table 4.1 summarizes past and probable future changes in extreme events. In the light of these projections and observed increases in costs relating to weather-climate hazards, policies and programs are necessary both to mitigate climate change (e.g., through emission reductions) and to adapt to its inevitable effects. Such adaptation is "the adjustment in natural or human systems in response to actual or expected climatic stimuli or their effects which moderates harm or exploits beneficial opportunities" (Parry et al. 2007, 27).

This type of adaptive capacity, or "the ability of a system to adjust to climate change (including climate variability and extremes) to moderate potential damages, to take advantage of opportunities, or to cope with the consequences" (IPCC 2007b, Appendix I), is determined by local or regional socioeconomic conditions. If the adaptive capacity can be enhanced, then there will be better means to cope with climate change. The key features of climate change for vulnerability and adaptation relate to variability and extremes, not simply to changing average conditions.

A comprehensive approach to climate policy should include long-term strategies both to reduce emissions (and hence meet the objectives of the UN Framework Convention and the Copenhagen Accord) and to manage the economic and social impact of climate change. Moreover, the strategy for adaptation needs to include plans to assist individuals and organizations in reducing their vulnerability to the deleterious effects of climate variability and change and also in capturing benefits from changes in weather and climatic patterns. The autumn 2010 report of Canada's Commissioner for Environment and Sustainable Development (CESD 2010) concluded: "The government has not established clear priorities for addressing the need to adapt to a changing climate. Although the government committed itself in 2007 to produce a federal adaptation policy to assist it in establishing priorities for future action, there is still no federal adaptation policy, strategy, or action plan in place. Departments therefore lack the necessary central direction for prioritizing and coordinating their efforts to develop more effective and efficient ways of managing climate change risks."

In the policy regime of emission reductions, the government's policy has varied. Announced in 2006 and maintained until the autumn of 2009, its target had been reducing Canada's total emissions of greenhouse gases by 20 per cent from 2006 levels by 2020 and by 60–70 per cent by 2050.[9] The first target corresponds to about 3 per cent reduction in comparison to the internationally agreed reference year of 1990. On 30 January 2010, as part of its commitment under the Copenhagen Accord, the government announced a new target of 17 per cent reduction from 2005 levels – the same as the United States. This goal is weaker than the earlier one and, if implemented, would increase emissions in 2020 by about 2.5 per cent relative to 1990 levels.[10]

INTERNATIONAL DEVELOPMENT

In 2002, participants at the World Summit on Sustainable Development (WSSD) in Johannesburg, South Africa, adopted a Summit Plan of Implementation as part of the strategy to meet the MDGs (Report of the World Summit for Sustainable Development 2002). The signatories agreed on a series of actions, which included protecting and managing the natural resource base of economic and social development. The report that followed the summit drew strong connections between international development and natural hazards and called for action:

> Article 37. An integrated, multi-hazard, inclusive approach to address vulnerability, risk assessment and disaster management, including prevention, mitigation, preparedness, response and recovery, is an essential element of a safer world in the twenty-first century. Actions are required at all levels to:
> (a) Strengthen the role of the International Strategy for Disaster Reduction and encourage the international community to provide the necessary financial resources to its Trust Fund ...
> (e) Improve techniques and methodologies for assessing the effects of climate change, and encourage the continuing assessment of those adverse effects by the Intergovernmental Panel on Climate Change ...
> (h) Develop and strengthen early warning systems and information networks in disaster management, consistent with the International Strategy for Disaster Reduction ...
> (j) Promote cooperation for the prevention and mitigation of, preparedness for, response to and recovery from major technological and other disasters with an adverse impact on the environment in order to enhance the capabilities of affected countries to cope with such situations.

Furthermore, the report drew connections between international development and climate change, which also demanded action: "38. Change in the Earth's climate and its adverse affects are a common concern of humankind. We remain deeply concerned that all countries, particularly developing countries, including the least developed countries and small island developing States, face increased risks of negative impacts of climate change and recognize that, in this context, the problems of poverty, land degradation, access to water and

food and human health remain at the centre of global attention ...
Actions at all levels are required to: (a) Meet all the commitments
and obligations under the [UN Framework Convention] ... " Both
this report and the Millennium Summit Declaration demonstrate
explicit connections between natural hazards, climate change, and
the MDGs.

Recognizing the impact of natural disasters on development, the
World Conference on Disaster Reduction of 2005[11] declared: " 1. We
will build upon relevant international commitments and frameworks,
as well as internationally agreed development goals, including those
contained in the Millennium Declaration, to strengthen global disaster reduction activities for the twenty-first century. Disasters have a
tremendous detrimental impact on efforts at all levels to eradicate
global poverty; the impact of disasters remains a significant challenge
to sustainable development. 2. We recognize the intrinsic relationship
between disaster reduction, sustainable development and poverty
eradication, among others, and the importance of involving all stakeholders." Again, an explicit link between disaster reduction and the
achievement of international development goals.

Canada's commitments to reducing disaster risk, addressing climate change, and furthering international development have come
through in a variety of statements. The Canadian International
Development Agency (CIDA),[12] which has the leading role in international development, has three priorities: increasing food security;
securing the future of children and youth; and stimulating sustainable economic growth. Prime Minister Harper announced, following the Muskoka Summit in June 2010, that the total Canadian
contribution for Maternal, Newborn and Child Health would be
$2.85 billion over five years.[13] A Sustainable Economic Growth
Strategy was announced by the minister of international cooperation on 25 October 2010. In 2009, according to OECD statistics,[14]
Canada tied with Austria with 0.30 per cent of GDP going for over-
seas development assistance, which made it 14th out of 23 countries.
This was a decrease of almost 10 per cent in real terms from 2008.
Canada's percentage was about a third or less of figures for Sweden,
Norway, Luxembourg, Denmark, and the Netherlands but higher
than the 0.21 per cent for the United States, which had increased its
contribution by about 6 per cent over 2008.

In climate change, Canada[15] has announced financing to support
developing countries' efforts to reduce greenhouse gas emissions and

adapt to the damage caused by climate change, with a focus on three areas – adaptation, clean energy, and forests and agriculture.

THE INTERSECTION OF ISSUES
AND BARRIERS TO CHANGE

Because of conflicting agendas and political pressures, it is difficult for governments to develop programs across a wide range of themes in a consistent way. One approach towards integration is the concept of sustainable development, which was first discussed in the 1980s. Beginning in the 1990s, many governments adopted sustainable development as a framework for reconciling economic development and environmental conservation, i.e., "to ensure that it meets the needs of the present without compromising the ability of future generations to meet their own needs" (World Commission on Environmental Development 1987, 8). In 1996, the government of Canada amended the Auditor General Act to create a Commissioner for the Environment and Sustainable Development (Office of Auditor General). The commissioner reports to Parliament on the government's efforts to protect the environment and foster sustainable development.

The Canadian International Policy Review (CIPR) (2005) outlined a strategy "for maximizing the contribution made by Canada and Canadians toward a world which is safer, healthier, more prosperous and more equitable; a world where development gains are sustainable," linking international development and sustainable development. As well, "Security and development are inextricably linked." Again, the security referred to implies terrorism (CIPR, 1), but there is a need to understand and redefine security in the broader sense to include natural hazards, as they are a bigger risk in most of the world. The review seemed more inclusive and integrative than most governmental approaches: "Government will: fully integrate development cooperation into Canada's international policy framework; ensure coherence across aid and non-aid policies that impact development" (CIPR, 7).

This chapter deals with three policy issues: disaster management, climate change, and international development. Each is linked to the other and also to sustainable development. In each case, there is also an element of uncertainty and risk – when will the next disaster happen?; how will the climate actually change?; and will other events, such as war or revolution, overwhelm international development? These issues have to be considered in the context of risk management. Governments need to bring these issues together in coherent and integrated policy development and implementation.

There is evidence that this convergence or intersection is starting to happen, but the "silos" (sectoral separation) of government agencies work against it. There is a clear need for leadership to make it happen for the benefit of everyone, in this generation and the ones to follow.

ACKNOWLEDGMENTS

The author appreciates discussions with and assistance from his colleagues in the Institute for Catastrophic Loss Reduction (ICLR) and at the University of Western Ontario. Preparation of the chapter was also influenced by discussions with many other people in Canada and around the world.

NOTES

1 Copenhagen Accord; www.unfccc.int
2 Cancun Agreement – draft decision – /CP.16. Outcome of the work of the Ad Hoc Working Group on long-term Cooperative Action under the Convention; www.unfccc.int
3 Speech from the Throne to Open the Third Session of the Fortieth Parliament of Canada. 3 March 2010. "Now and for the Future." ISBN 978–1–100–14874–8; cat. no. SO1–1/20109E-PDF, ISSN 1493–3551. Available at www.pco-bcp.gc.ca
4 UN International Strategy for Disaster Reduction (ISDR), www.unisdr.org
5 www.publicsafety.gc.ca/prg/em/ndms/drr-eng.aspx
6 UN International Strategy for Disaster Reduction (ISDR). www.unisdr.org/terminology
7 www.publicsafety.gc.ca/prg/em/miti-eng.aspx
8 www.unfccc.int
9 climatechange.gc.ca/default.asp?lang=En&n=72F16A84–1 (accessed 16 Jan 2010).
10 Shawn McCarthy, *Globe and Mail*, 27 Jan. 2010: "Tories hedge on emissions targets."
11 UN ISDR. www.unisdr.org/eng/hfa/hfa.htm
12 www.acdi-cida.gc.ca/acdi-cida/ACDI-CIDA.nsf/eng/HEL-1027152651–QTD #tphp

13 www.pm.gc.ca/eng/media.asp category=1&featureId=6&pageId=26&
id=3479
14 www.oecd.org/document/9/0,3343, n_2649_37413_1893129_1_1_1_
37413,00.html
15 www.climatechange.gc.ca/default.asp?lang=En&n=5F50D3E9-1

REFERENCES

Berke, P.R., Y. Song, and M. Stevens. 2008. "Integrating hazard mitigation
into new urban and conventional developments." *Journal of Planning
Education and Research* 28: 441–55.
Borenstein, S., and J. Reed Bel. 2010. "2010: deadliest for natural disasters
in a generation." *Globe and Mail*, 19 Dec.
Canadian International Policy Review (CIPR). 2005. "Overview: Canada's
International Policy Statement." A Role of Pride and Influence in the
World. International Policy at a Crossroads. Archived document
www.international.gc.ca/cip-pic/documents/IPS-EPI/international-
internationale.aspx?lang=eng
Canton, L.G. 2007. *Emergency Management: Concepts and Strategies for
Effective Programs*. Hoboken, NJ: Wiley Inter-Science, 2007.
CESD (Commissioner of the Environment and Sustainable Development).
2010. "The fall 2010 report of the Commissioner of the Environment and
Sustainable Development to the House of Commons." www.oag-bvg.gc.ca
Copenhagen Diagnosis. I. Allison et al., eds. 2009. *Updating the World on
the Latest Climate Science*. University of New South Wales Climate
Change Research Centre (CCRC), Sydney, Australia: 60.
Cullen, R. 2004. Speaking Notes for the Honourable Roy Cullen, Parliamentary Secretary to the Minister of Public Safety and Emergency Pre-
paredness Canada, Presentation to 1st Annual Symposium of CRHNet,
18 Nov.
Gall, M., K.A Borden, and S.L. Cutter. 2009. "When do losses count? Six
fallacies of natural hazards loss data." *Bulletin of the American Meteo-
rological Society* 90: 799–809.
Godschalk, D.R. 1991. "Disaster mitigation and hazard management." In
T.E. Drabek and G.J. Hoetmer, eds., *Emergency Management: Principles
and Practice for Local Government*. Washington, DC: International City
Management Association: 131–59.
Godschalk, D.R., T. Beatley, P. Berke, D. Brower, and E.J. Kaiser. 1999.
Natural Hazard Mitigation. Washington, DC: Island Press.
Handmer, John, 2003. "Adaptive capacity: what does it mean in the con-
text of natural hazards?" In Joel B. Smith, Richard T. Klein, and

Saleemul Huq, eds., *Climate Change, Adaptive Capacity and Development*. London: Imperial College Press: 51–70.

Henstra, D., and G. McBean. 2005. "Canadian disaster management policy: moving toward a paradigm shift?" *Canadian Public Policy* 31(3): 303–18.

IPCC (Intergovernmental Panel on Climate Change). 2007a. Core Writing Team: R.K. Pachauri and A. Reisinger, eds., *Climate Change 2007: Synthesis Report. Contribution of Working Groups I, II and III to the Fourth Assessment Report of the Intergovernmental Panel on Climate Change*. Geneva: IPCC: 104.

– 2007b. M.L. Parry, O.F. Canziani, J.P. Palutikof, P.J. van der Linden, and C.E. Hanson, eds., *Climate Change 2007: Impacts, Adaptation and Vulnerability. Contribution of Working Group II to the Fourth Assessment Report of the Intergovernmental Panel on Climate Change*. Cambridge: Cambridge University Press, Appendix I: Glossary: 976.

Matson, P., et al. 2010. *Advancing the Science of Climate Change*. Washington, DC: National Academies Press. www.nap.edu

Mileti, D.S. 1999. *Disasters by Design: A Reassessment of Natural Hazards in the United States*. Washington, DC: Joseph Henry Press.

Millennium Declaration and Goals. 2008. www.un.org/ millenniumgoals/

Mirza, M.M.Q. 2003. "Climate change and extreme weather events: can developing countries adapt?" *Climate Policy* 3: 233–48.

Munich Re. Group. 2005. As reported in the *WCDR News*, published by the Secretariat of the Hyogog Cooperative Committee for the World Conference on Disaster Reduction 4: 21 Jan.

– 2008. Topics Geo-Natural catastrophes 2008 Analyses, assessments, positions. 2008–302–06022–en.pdf.www.munichre.com/touch/ publications/en/list/default.aspx?category=17

– 2010a. Media statement. Number of weather extremes a strong indicator of climate change, 8 Nov. 2010. www.munichre.com/en/media_relations/company_news/2010/2010–11–08_company_news.aspx

– 2010b. Topics Geo: Natural catastrophes 2009 Analyses, assessments, positions. 2010–302–06295–en.pdf. www.munichre.com/touch/ publications/en/list/default.aspx?category=17

Mutter, John C. 2005. "The earth sciences, human well-being, and the reduction of global poverty." *EOS, Transactions American Geophysical Union* 86 (16): 157, 164–5.

National Research Council (Britain). 2010. "Report in brief: advancing the science of climate change: America's climate choices." *United Kingdom's National Security Strategy* (2008): 4. americasclimatechoices.org

Parry, M.L., O.F. Canziani, J.P. Palutikof, et al. 2007. "Technical summary." In M.L. Parry, O.F. Canziani, J.P. Palutikof, P.J. van der Linden, and C.E.Hanson, eds., *Climate Change 2007: Impacts, Adaptation and Vulnerability. Contribution of Working Group II to the Fourth Assessment Report of the Intergovernmental Panel on Climate Change.* Cambridge: Cambridge University Press: 23–78.

Prater, C.S., and M.K. Lindell. 2000. "Politics of hazard mitigation." *Natural Hazards Review* 1(2): 73–82.

Report of the World Summit for Sustainable Development. 2002. Johannesburg, South Africa, 26 Aug–4 Sept. A/CONF.199/20*. www.un.org

Rodriguez, J., F. Vos, R. Below, and D. Guha-Sapoir. 2009. *Annual Disaster Statistical Review 2008: The Numbers and Trends.* Brussels: Centre for Research on the Epidemiology of Disasters. www.emdat.be

Solomon, S., D. Qin, M. Manning, et al. 2007. "Technical summary." In S. Solomon, D. Qin, M. Manning, Z. Chen, M. Marquis, K.B. Averyt, M. Tignor, and H.L. Miller, eds., *Climate Change 2007: The Physical Science Basis. Contribution of Working Group I to the Fourth Assessment Report of the Intergovernmental Panel on Climate Change.* Cambridge and New York: Cambridge University Press: Table TS4: 52.

UNISDR (United Nations International Strategy for Disaster Reduction). *Global Assessment Report on Disaster Risk Reduction.* Geneva: United Nations: 207. www.preventionweb.net/english/hyogo/gar/report/index.php?id=1130&pid:34&pih:2

United Kingdom's National Security Strategy. 2008. *The National Security Strategy of the United Kingdom: Security in an Interdependent World.* Presented to Parliament by the Prime Minister, by command of Her Majesty. March, Cm 7291: 64. www.interactive.cabinetoffice.gov.uk/documents/security/national_security_strategy.pdf

Vos, F., J. Rodriguez, R. Below, and D. Guha-Sapir. 2010. *Annual Disaster Statistical Review 2009: The Numbers and Trends.* Brussels: CRED. www.cred.be

Wahlström, M. 2009. Assistant Secretary-General for Disaster Risk Reduction and Special Representative of the UN Secretary-General for the Implementation of the Hyogo Framework for Action, quoted in J. Birkmann, G. Tetzlaff, and K.O. Zentel, eds., *Addressing the Challenge: Recommendations and Quality Criteria for Linking Disaster Risk Reduction and Adaptation to Climate Change.* DKKV Publication Series 38: 5.

World Commission on Environment and Development. 1987. *Our Common Future*: 8. www.un-documents.net/wced-ocf.htm

5

Mountain Hazards and the Resilience of Social-Ecological Systems: Examples from India and Canada

JAMES S. GARDNER AND JULIE DEKENS

INTRODUCTION

Until recently, perceptions of mountains often evoked danger (Nicholson 1963). Nevertheless, for generations people have inhabited mountains, living with dangers posed by earthquakes, landslides, avalanches, flash floods, fires, cold temperatures, storms, wild animals, and so on (Hewitt 1997a). Any process or condition that constitutes a threat to human safety and property may be considered a hazard. Today, perceptions of danger in mountains have eased, record numbers of people travel through, visit, and inhabit mountain regions, and levels of risk and vulnerability are still high, if not increasing (Hewitt 1997a). During the past three decades, stresses on the physical and biological systems of mountain ranges have intensified many fold as the world's population has doubled and mountain regions' total population has more than tripled (Slaymaker 2010). Even if the frequencies and magnitudes of hazard processes are not increasing in these areas, the levels of risk and vulnerability are, because of economic and social changes (Gardner 2002).

Three factors are invaluable in understanding hazards in mountain regions. First, these areas are relatively active geophysically and hydrologically, and they are biologically diverse because of an altitude- and aspect-driven variability in energy and moisture. Second, they are diverse in the make-up of their social systems, which range from relatively small, isolated settlements based on subsistence agriculture, animal husbandry, and/or gathering and hunting to complex, diversified, and linked population centres made up of permanent residents, economic migrants, amenity migrants, tourists, and other transients. Third, their links with other areas are defined by flows of air, water, materials, animals, people, goods, services, information, money, and influence or authority, and in every respect these connections with lowlands have increased in number and importance.

The understanding of hazards and consequent disasters rests as much on knowledge of the human dimensions as on the bio-geophysical dimensions (Steinberg 2000; Klinenberg 2002; Wisner et al. 2004; Bankoff, Frerks, and Hilhorst 2007). The work of Barrows (1923) and early research on natural hazards, exemplified in Burton, Kates, and White (1978) and Hewitt and Burton (1971), recognized the role of both natural and human factors. Yet much subsequent research and mitigation have focused on forecasting, controlling, and/or preventing bio-geophysical conditions and processes, while neglecting human factors and doing little to reduce loss of life, injury, property damage, and disruption of economic or other activity (Hewitt 1983; Haddow, Bullock, and Coppola 2008). Mountain areas have become increasingly disaster-prone and experience disproportionately more disasters than other environments (Hewitt 1997b).

The goal of this chapter is to describe hazards in mountains social-ecological systems through the concept of resilience (Folke, Colding, and Berkes 2003; Resilience Alliance 2003). Social-ecological systems are integrated systems of people, including their resource-use practices and technological and institutional arrangements, set within their natural environments. We hypothesize that resilience-building and enhancement in these systems can ameliorate and mitigate the impact of hazardous processes. Resilience-building and enhancement in this context refer to increasing the ability of community members to learn and adjust, to use all forms of knowledge, to self-organize, and to develop constructive institutional links with other and higher-order social-ecological systems. The objective of the chapter is to articulate the conditions that contribute to resilience-building and enhancement in these systems in the face of hazards and in the aftermath of a damaging and destructive event (i.e., a disaster).

The method of the chapter is expository and historical. That is, examples illustrate a concept – resilience – and the elements that support it (or do not) in a mountain environment. The examples are not meant to be representative of all cases. The data and information come from published (i.e., secondary) sources and archival records (i.e., unpublished reports, diaries, letters, newspapers, maps, and photographs). The examples illustrate similarities and differences between situations in contrasting societies. Most important, they provide a historical dimension. Indeed, in the best of all worlds, the past teaches people. History also shows where learning does not take place or where conditions leading to, and accompanying, disaster are immutable through time. In either case, this chapter presents illustrations to emphasize the value of learning from disasters, particularly if the goal is to make social-ecological systems more resilient and sustainable. Finally, the examples illustrate similarities and differences between hazards of varying types.

RESILIENCE AS A CONCEPT

Resilience has emerged in literature on psychology, ecology, food aid and famine, resource management, health, and climate change (e.g., Holling 1986; Berkes, Colding, and Folke 2003; Folke, Colding, and Berkes 2003; Bingeman, Berkes, and Gardner 2004; Chapin et al. 2004; Norris et al. 2008). Some limited applications appear in recent treatments of hazards in mountain regions (Robledo, Fischler, and Patino 2004; Weichselgartner and Sendzimir 2004). Resilience is sometimes presented as a corollary of vulnerability (Buckle, Mars, and Smale 2000), and both terms have been used with a variety of meanings and without consensus. Kasperson and Kasperson (2001) define resilience as a component of vulnerability, and Buckle, Mars, and Smale (2000) point out that qualities of a community that reduce vulnerability are indicative of resilience.

However, the relationships between vulnerability, as the degree of exposure of a social-ecological system to hazards and risks (Wisner et al. 2004), and resilience may be more complex. Each may have positive and negative correlates, depending on the situation. A social-ecological system that is highly resilient at a certain location or time also may be highly vulnerable, and vice versa. Much of the hazards literature focuses on vulnerability. The focus on resilience-building is a positive approach that emphasizes the strengths of a system as opposed to its weaknesses.

Berkes, Colding, and Folke (2003) identify resilient social-ecological systems as those that sustain livelihoods in the face of change through self-organization, re-organization, and learning. Characteristics that enhance resilience include: vibrant leadership, shared goals and values, established institutions and organizations, positive socio-economic trends (stable and healthy population and diversified economic base), constructive external partnerships and links, and availability and use of resources and skills (Buckle, Mars, and Smale 2000) (Figure 5.1). As such, they will be sensitive to locational and temporal aspects of hazardous processes and will seek to incorporate diversity, redundancies, skills, resources, technologies, partnerships, and institutions to mitigate the impact of hazards. Among these characteristics, those relating to human and social capital are key to building and enhancing resilience. Social capital includes the capacity of individuals and groups to build relations of trust and reciprocity, to adhere to commonly agreed rules, norms, and sanctions, and to be able to work together and with other institutions (e.g., Ostrom 1992; Pretty and Ward 2004). Social capital in the form of associations and entrepreneurial behaviour can influence the degree of cooperation locally and beyond and thus affect people's ability to adjust in the face of change.

Yet the same characteristics that enhance resilience can lead to rigidity and conservatism in the face of stresses, strains, and crises. For instance, shared goals and values and established institutions and organizations may produce resistance to change so that the system becomes less resilient in the face of future hazards. The corollary of each of these characteristics almost certainly produces high vulnerability and inability to cope and adapt in the context of disaster (Dekens 2005).

The key qualities in building resilience are the ability to learn, i.e., to acquire knowledge, and the ability to apply it to a situation or anticipated situations, i.e., to adapt. Knowledge is acquired through experience and observation, including those of other people past and present, and through the ability to apply general principles to particular situations. Adaptation requires using the knowledge to purposefully adjust the social-ecological system and/or the physical environment – in this case, to reduce the impact of future hazardous events, including those that may be unanticipated. Thus adaptation or adaptability is central to resilience. Much of the research on human adaptation in a social-ecological context is described in the literature of geography and anthropology (e.g., Firth 1969), and the focus has been on local realities and the static measures in place

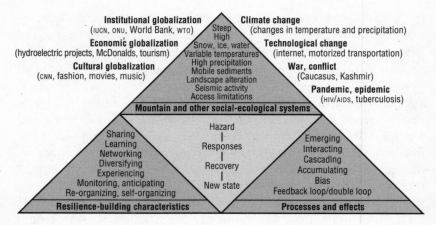

Figure 5.1
Building resilience of mountain social-ecological systems

Note: The impacts of any particular hazard process (i.e., hazard, responses, recovery, new state) are a product of the magnitude of a biogeophysical process and the levels of vulnerability and resilience of the social-ecological system. These are influenced by global factors, different types of processes and effects, and the resilience-building characteristics of the social-ecological system.

rather than on the dynamics of change (Goodman and Leatherman 2001). Batterbury and Forsyth (1999) point out that there are many more components to sustaining livelihoods than simple adaptation to environmental change, and these components are being recognized in complex systems theory and other pertinent literature (e.g., *Focus* 2009). Limited attention to cross-scale effects and influences and their static nature (Brooke 2001) has led to criticism of the human adaptation approach and attempts to move beyond it to the concept of resilience.

RESILIENCE AND SCALE

Most hazards and disasters have local and external components at different scales. Interacting and cascading effects operate within and across scales and make the understanding and mitigation of hazards complicated. A particular landslide or flood may test the resilience of a social-ecological system, but factors far removed from the time and location of the event also may influence resilience – for instance, global factors, including climate change, technological change, economic, cultural, and institutional globalization, war and conflict, and pandemics and epidemics (Table 5.1 and Figure 5.1). Their influences are eventually felt even in the most remote mountain communities, and the rapidity of their impact has increased over time.

Each global factor may have particular outcomes at the regional/national level that in turn influence resilience and the impact of hazard events at the local level. For example, global climate change may magnify regional shifts in temperatures and precipitation. Archaeological and historical records bear witness to movements of human populations into and out of mountain areas coincident with climate variations over the past 40,000 years. Continuing climate change and variation in mountain areas could lead to changes in the frequency and magnitude of floods, fires, landslides, snow avalanches, droughts, and so on, thus influencing hazards at the local scale. Various global economic changes, such as the Great Depression of the 1930s, and pandemics, such as the "Spanish" influenza of 1918, affected local mountain communities, increasing their vulnerability and decreasing resilience in the face of further shocks, stresses, or crises.

Interacting and cascading effects operate across scales. For example, a mountain social-ecological system may become less resilient as a result of an ageing population, which itself may result from out-migration or the diminution of a younger, productive segment of the population. In turn, this out-migration may be due to the external attraction of cash-paying jobs, disturbances such as war, and diseases such as HIV/AIDS. In another example, a mountain social-ecological system may become less resilient to hazards associated with unusual, high-magnitude rainfall events that produce catastrophic soil erosion and loss of agricultural productivity. This may result from deforestation in the surrounding area. The deforestation may have been stimulated by regional or global increases in demand and therefore in prices for wood products and facilitated by the incursion of a sophisticated commercial system of contract buying, as occurred in parts of the Himalaya under British administration in the nineteenth century (Tucker 1982; Wasson et al. 2008).

Cascading effects in hazard events, often operating across scales, have tested and eroded the resilience of mountain social-ecological systems. For example, in high mountain regions such as the Himalaya and Andes, earthquakes are usually regional in their impact, causing the collapse of structures and damage to infrastructure of all sorts over wide areas (Hewitt 1976). At the same time, they may produce secondary landslides and snow avalanches that

Table 5.1
Global factors affecting hazards and resilience in mountain regions

Factor/trend	Mechanisms (examples)	Local impacts (examples)
Climate change	Reduction/increase in precipitation and/or temperature	Increased/decreased prevalence of floods, droughts, landslides, storms, avalanches, fires
	Reduction/increase in precipitation type and intensity	Change in seasonality or timing of weather thresholds such as onset of frosts, onset of monsoon/rainy season
	Reduction/increase in extreme temperatures	Change in growing season, degree-days, and so on
Globalization	Dispersal and diffusion of information, ideas, preferences, materials, technologies, and so on	Increased demand for new goods and services (mountain tourism)
		Raised expectations vis-à-vis living standards
	Changes in market conditions	New markets for local products (medicinal plants)
	Changes in number and type of employment opportunities	Increased cash income and demand for cash income (out-migration)
		Greater competition for markets (local products no longer competitive)
War/conflict	Restrictions and limitations on movement and travel	Population decline
	Changes in political and social relations and alliances	Death and injury in civilian population (loss of productive capacity)
	Changes in population and its structure	Erosion of trade networks and economic base (tourism)
	Forced migrations	Destruction of property and infrastructure
		Damage to environment and resource base (soil and water contamination)
		Depopulation of specific areas/regions
Organizational shifts	Changes in international/national protocols, standards, and alliances	Conservation protocols leading to land use restrictions and limitations (protected areas)
	Changes in political and government structures	New livelihood opportunities (eco-tourism)
		New governmental priorities (focus on environmental protection, basic education, health care)
Pandemics/epidemics	Population reductions	Loss of productive capacity
	Changes in population age structure and sex ratios	Isolation and disruption of economic and social links (tourism and trade)
	Chronic illness and reduction in productive capacity	Resources diverted from livelihood support (economic decline)
	Reductions in life expectancy and longevity	
Improved health care, nutrition, and so on	Increase in population	Agricultural expansion, deforestation, erosion, and so on
	Increase in demand for food and services	Increased waste production, pollution, and contamination
	Increased longevity and life expectancy	Productive ratio of capacity to population may decrease or increase.

have their own more localized effects, including damage to structures and roads and death and injury to people (Barnard et al. 2001; Slaymaker 2010). Hewitt (1984 and 1997b) suggests that the impact of slope failures arising from earthquakes has been underestimated. One of the inherent vulnerabilities in mountain terrain is that access is especially susceptible to blockage by landslides and snow avalanches. The linearity of power transmission lines, pipelines, and land-based telephone lines also makes them especially vulnerable. The disruption and interruption of access, communications, and energy and water supplies significantly hamper rescue and recovery and delay medical and other services, which may engender the spread of infectious diseases, which in this case is a tertiary hazard. Thus an earthquake produces a number of disruptions across scales through a cascade of subsidiary hazardous processes and effects, thus testing the resilience of social-ecological systems in several ways.

Finally, hazards and disasters do not have entirely negative consequences. Resilient social-ecological systems may benefit from new opportunities presented in the aftermath. First, these may occur as people learn from the crisis and make adjustments to deal with future events. Second, a disaster may produce an infusion of large sums of money and other resources through institutional cross-scale links. Such relief is usually for rebuilding infrastructure and housing, which generates new jobs and other economic opportunities through spin-off effects, in addition to replacing ageing and inadequate infrastructure that could have otherwise been a drain on the social-ecological system. Third, the infusion of emergency medical and social services may produce a legacy through the continuation of a higher level of such services on a permanent basis. Fourth, some hazardous processes result in the revitalization of resources such as soil and plant life, as in the cases of floods, hurricanes, wildfires, and volcanic eruptions (e.g., Colding, Elmqvist, and Olsson 2003; Chapin et al. 2008). Finally, disasters make the news, and the affected social-ecological systems may be able to turn this notoriety and attention to their future benefit through the development of "attractions" for visitors and aid from international donors. Thus resilience may be measured as well by the ability of a social-ecological system to take advantage of opportunities presented by hazards and disasters. At the same time, cross-scale dependencies may be created that discourage local initiatives to build resilience.

BUILDING RESILIENCE

Building resilience is necessary for the effective mitigation of hazards. Folke, Colding, and Berkes (2003) identify learning from crises, nurturing diversity, using all forms of knowledge, and creating opportunities for self-organization as attributes of resilience-building and enhancement. Analysis below shows that developing external links and partnerships may be important as well. Building resilience is a challenge in the rapidly changing conditions in many mountain regions. Changes include shifts in bio-geophysical conditions, expansion of infrastructure such as buildings, facilities, and roads, erosion of traditional knowledge and practices, natural population growth through reduction in mortality due to improved nutrition and health care, in-migration of permanent and transient residents, natural resource extraction, development of commercial agriculture and horticulture, protection of strategic interests, national security, war, and tourism development (e.g., Gardner et al. 2002).

Learning

Learning from previous crises, including those elsewhere, can help in building resilience and mitigating the impact of hazards. Mountain regions provide examples of such learning as well as of situations where people have not learned from experience (de Scally and Gardner 1994). Experience is useful in identifying locational and temporal aspects and conditions associated with event occurrence. Many mountain communities know through generations the areas subject to floods, landslides, and avalanches and the conditions under which they occur. They know how to avoid dangerous areas and restrict their use at dangerous times. Today, sophisticated knowledge and technologies exist for hazard assessment and mapping that can assist in this process (Gardner and Saczuk 2004). The apparent absence of learning may be related to limited alternatives; lacking, ignored, or lost knowledge; inadequate transfer of knowledge between places, communities, and organizations; and inadequate tradeoffs between the perceived benefits of the status quo and the costs of changing.

Diversity

Social-ecological diversity builds resilience by creating and maintaining options and choices. This is especially important in maintaining sustainable livelihoods. Learning how and what diversity to enhance can come about through experience. Various levels and types of ecological diversity are present in mountain regions, as demonstrated by topographical, geological, climatic, hydrological, and biological diversity within relatively small areas. The range in altitude and slope aspect produces variable moisture and energy balances that in turn generate a diversity of micro-climates and ecosystems. The resulting diversity of natural products and conditions can support a variety of livelihood activities. Typically, mountain social-ecological systems rely on a range of activities that includes hunting, gathering, shifting and sedentary agriculture, animal husbandry, and horticulture. Within the individual activities, diversity and redundancy may be built in to protect against hazards. Diverse crops provide dietary variety as well as protection against damage by species-specific diseases. Household fields may be dispersed rather than contiguous to provide protection in the event of floods, water shortages, landslides, and diseases that are place specific. Climatic and ecological seasonality is a form of temporal diversity. Although hazards and limitations accompany seasonality, such seasonal variation also increases the variety of livelihood options. Examples are found in the lower-elevation Himalaya, where winter and summer cropping alternates wheat, barley, and maize with paddy rice, and in temperate mountains, where winter ski tourism replaces other forms of summer tourism and agriculture.

Social-ecological diversity may be enhanced further by manifestations of modernity. Economic activities that accompany incursions of people, facilities, and infrastructure into the mountains provide cash employment that can supplement or supersede traditional practices. Tourism development is an example. Such development opportunities increase income and diversify livelihood options that, in combination, enhance resilience. However, over-reliance on new opportunities, such as tourism, without attention to diversity may increase vulnerability to hazards. The collapse of the tourist industry and livelihoods in Kashmir with the rise of armed conflict in 1989 (Gardner et al. 2002) is an example, as is the decline of tourism throughout the mountains of South Asia following the terrorist attacks on the United States of 11 September 2001. Further, the process of economic diversification does not mean that everyone in a social-ecological system can benefit from it. Lack of equality and equity within such systems may prevent or erode resilience and increase vulnerability among the disadvantaged populations.

Local Knowledge

Davidson-Hunt and Berkes (2003), Folke, Colding, and Berkes (2003), and Berkes (2007) demonstrate that combining different types and systems of knowledge can enhance resilience. Lack of attention and sensitivity to local knowledge can lead to increased vulnerability of life and property. Social-ecological systems present in hazardous environments over many generations usually have resilience in many forms that are not always apparent. In addition, they may have customary land use practices, which are designed to ensure livelihoods. The associated knowledge, often referred to as indigenous, traditional, or local knowledge, is contained and transmitted in forms different from those associated with Western scientific knowledge (Ramakrishnan et al. 2005). As a result, outside interests may discount, discredit, or simply ignore the indigenous knowledge. In the context of resilience, the resulting dangers are threefold: innovators may improperly place or time new facilities, people, and activities; new technologies may increase vulnerability; and new institutions, rules, and procedures may render local livelihoods less sustainable, making the people and communities less resilient.

For the most part, members of indigenous social-ecological systems in the mountains have an acute understanding of environmental resources and hazards (Duffield et al. 1998; Joshi and Singh 2010). The results are seen in the location, materials, and building styles of homes and other structures, the layout of settlements, the organization and types of agricultural practices and products, and the annular patterns of life. Lack of attention by newcomers to the knowledge and understanding that underlie these arrangements can result in unnecessary risk and sometimes disaster (de Scally and Gardner 1994).

Interventions may change or disrupt local practices and make people less resilient. A common example comes about as a result of new land use or land tenure rules. The establishment of Reserved Forests in the mountainous parts of British India in the nineteenth century altered indigenous livelihood systems and made some people and communities less resilient by denying them access to forest-based livelihoods. The establishment of parks, wildlife reserves, and other protected areas in mountain regions following international standards and guidelines has eroded traditional livelihoods at the local level in Kunjerab Park in northern Pakistan (Ali and Butz 2003),

Great Himalayan National Park in Himachal Pradesh (Saberwal and Chattre 2001), and Nanda Devi Biosphere Reserve in Garhwal, for example. An extreme case is found in the story of the Ik (mountain) people of northwest Kenya, whose social-ecological system was destroyed by creation of a national park in combination with the onset of drought (Turnbull 1972). Building resilience in mountain social-ecological systems demands recognition of, and attention to, local knowledge and practices to avoid these situations.

Self-Organization

Self-organization refers to the ability of a social-ecological system to establish agencies, arrangements, and institutions to mitigate the effects of hazards. Self-organization is useful in forecasting and publicizing hazard events, reacting in an organized and effective way to an emergency, and having in place organizations and institutions to oversee longer-term rehabilitation and reconstruction. In traditional mountain settings, examples arise as a result of learning through experience in dealing with hazards and risks. The learning produces local knowledge, which is retained and passed on by a variety of means. The knowledge is translated into practical measures that may help mitigate and ameliorate hazards and risks and deal with the aftermath of disasters. Sharing learning and creating a common, intergenerational body of knowledge require self-organization. Community sanctions and prohibitions against building in dangerous areas are products of self-organization. Common methods of avoidance, such as the migration of whole communities out of danger zones at certain times of the year, as happens in snow- and avalanche-prone areas of the western Himalaya during winter (de Scally and Gardner 1994), offer other examples of self-organization that build resilience.

Assisted by the introduction of new technologies and knowledge, communities may use various means of self-organization to develop warning systems that ameliorate hazard and risk. Flood, landslide, and avalanche warning systems, including the technologies and administrative structures, are examples. More often, local knowledge and existing institutions are either ignored or overridden, and new laws and practices are put in place. In mountain areas subject to external influences, including the in-migration of new people and activities, the traditional prohibitions may become codified in law, or they may not, and then new practices and laws are codified to suit the newcomers. The establishment in the affected place of organizations, technologies, and institutions may or may not improve resilience in the face of hazards.

Links and Partnerships

The establishment of institutional links and partnerships is an essential component in the mitigation and amelioration of, and recovery from, disasters and therefore in resilience-building. The role of external influences and links has been twofold. First, external influences have contributed to the erosion of resilience in mountain regions through influences on bio-geophysical conditions and, more so, through alterations in the extent, intensity, and type of human activities, thus exposing more people and property to danger and, through social change, altering the ability of communities to adjust to disaster. Second, external links and partnerships do provide a medium through which resources and assistance flow to the affected area and populations in the event of disaster. Without this flow, the sustainability of some communities would be doubtful. We would argue that this is of critical importance in mountain regions where the physical reality of isolation from the main centres of population and authority has exacerbated disaster in the past. Where the balance lies is difficult to tell and varies from one situation to the next (Dekens 2005).

RESILIENCE IN MOUNTAIN SOCIAL-ECOLOGICAL SYSTEMS: EXAMPLES

Mountain regions provide many examples by which to examine the role of resilience in social-ecological systems in the face of hazards. As we noted above in our discussion of method, the examples are meant to be expository and not representative. They illustrate some key points, as we outlined above. By using a historical approach, we can demonstrate lessons learned and not learned. Some factors leading to disaster may be immutable and unchangeable through time. Not only have the levels of risk and the impacts of hazards in mountain regions increased over the past century, largely as a result of human activities, but the possibilities and means through which to ensure resilience of mountain social-ecological systems have increased as well. Clearly, this has happened in some instances but not in all.

Epidemics in the Kootenays, British Columbia, Canada

Epidemics of new contagious diseases (smallpox, measles, influenza) swept through the southern interior mountain region of British Columbia in the nineteenth century (Harris 1997; Gordon 2004). In the Kootenays (50N/118W), the impact was sufficient to decimate the existing populations of indigenous peoples such that newly arrived migrants entered into essentially "uninhabited" lands. The legacy today is a confusion of traditional rights and codified ownership. This is a story repeated in many mountain areas of the world (Hewitt 1997b), and elsewhere, as outsiders introduced new diseases. In the nineteenth century and earlier, no effective means of treatment or immunization were available, and natural immunities were rare, if they existed at all. In the face of such hazards, human vulnerability was extremely high, and social-ecological systems collapsed quickly as their productive and reproductive capacities eroded. The situation in the nineteenth century in the Kootenays is an extreme example in the sense that there was no prior learning, local knowledge provided no effective responses, helpful external links were non-existent, and effective interventions did not exist anywhere. It is a historical example, but could it be repeated?

Infectious diseases continue to impose a toll on mountain social-ecological systems (Hewitt 1997b) despite a build-up of resilience through interventions for prevention, treatment, and care. Gastro-intestinal infections, cancers, tuberculosis, and influenza, some drug-resistant, continue to take a toll in loss of life and livelihoods. The emergence in the 1980s of HIV/AIDS as a pandemic that has reached into the mountain regions worsened the situation (Seddon 1995a; Cox 2000). Bio-medically, HIV is easily prevented and is becoming treatable. Its growth to pandemic proportions has been driven by social, economic, and political factors that are less easily controlled. Few aspects of this hazard are distinctive to mountain regions apart from the fact that men and women in mountain communities have become increasingly mobile in search of cash-paying jobs, and that mobility is a major factor in the spread of HIV infection (Mann 1992; Seddon 1995b). HIV's transformation to AIDS in individuals leaves them open to opportunistic infections, such as some of the above, which do the killing. Its impact is felt most among the productive age groups and the young in any social-ecological system, thereby attacking the very basis of its resilience. Young women are increasingly subject to infection through heterosexual transmission. The pandemic is a disaster in progress, and it could well be the defining human disaster of the twenty-first century, as were smallpox and measles for the indigenous mountain people of the Kootenays in the nineteenth century and the "Spanish" flu for the newcomers in the Kootenays and others the world over in 1918.

There are three lessons in this example:

- The vulnerability of human populations to new infectious diseases is extremely high, and this has not changed through time.
- The geographical and social isolation of some populations makes interventions, especially those enacted through external links, very difficult.
- The geographical, social, and economic characteristics of social-ecological systems strongly influence the impact of this hazard, and the impact is felt through generations.

Floods and Debris Torrents in Kullu, Himachal Pradesh, India

Floods and debris torrents are a common hazard in the Kullu District (32N/77E) of Himachal Pradesh, in the heart of the Pir Panjal Himalaya. They include monsoon rainfall and snowmelt floods on the main stem of the Beas River and rainfall-generated floods in its steep, low-order tributaries. Some speculation in the media and scientific literature suggests that the frequency and impact of these events increased during the twentieth century due to deforestation and overgrazing in this part of the Himalaya. The speculation arises in the aftermath of deadly and damaging events such as those in 1993, 1996, 2002, and 2003. However, subsequent research (Gardner 2002) has demonstrated that deforestation has been limited in the relevant areas, and any increase in the impact of the events can be attributable mainly to the extension and intensification of human activities in the area.

The geography of settlements in Kullu attracted the attention of early colonial administrators in the 1850s (Harcourt 1871). Each

settlement consisted of a tight assemblage of wood and stone structures, situated on valley-side terraces well away from the Beas River
and, for the most part, away from valley-side streams, gullies, and active landslide areas and avalanche slopes. Settlements were located at
approximately regular intervals. Each was surrounded by a layout of
cultivated and usually terraced fields owned by members of the community. This pattern was conditioned in part by local knowledge of
the most dangerous places with respect to flooding and related hazards. In other words, the permanent residents had developed some
measure of resilience through locational choices. This was buttressed
and supplemented by institutions (i.e., rules) about the building of
residences outside the confines of existing settlements.

A review of documentary records of the area from 1850 to the
present provides little evidence of flood damage until 1894, when a
major flood/debris torrent swept down the Phojal Nalla (stream),
killing over 200 people and hundreds of sheep, goats, and cattle and
destroying crops near its confluence with the Beas River (*Civil and
Military Gazette* 1894). A landslide dammed the stream and created
a small lake, followed by a dam burst and torrent that swept through
a temporary encampment of traders and herders on the active floodplain at the Beas/Phojal confluence. This is a demonstration of the
relatively higher vulnerability of temporary residents by virtue of
location, in contrast to permanent settlements. The loss of many
traders affected the longstanding trade route from Kullu to Lahul,
Spiti, and Tibet for years afterwards, giving advantage to the trans-
Himalayan trade links through Kashmir and Garhwal (*Civil and
Military Gazette* 1894; Rizvi 1999).

The 1993 event also affected the Phojal Nalla, destroying and
damaging new homes and shops outside the traditional settlement
areas. The events of 1996, 2002, and 2003 disproportionately affected camps of migrant construction workers and their families, in
this case from the plains and neighbouring Nepal. These losses did
not have long-term impacts as in 1894 because surplus and mobile
labour exists in South Asia today, and losses are quickly and easily
replaced.

There are four lessons in this example:

• The rapid onset of damaging events in the mountains affords little warning.

- Prior learning about hazardous terrain by long-established residents provides the basis for placing settlements and valuable property in safe locations.
- Newcomers, temporary residents, and migratory workers may occupy land that is highly vulnerable to hazards.
- Hazardous processes recur in the same locations, and recurring and increasing impacts at those locations relate as much to human as to physical factors.

The 1905 Earthquake in Kangra, Himachal Pradesh, India

The Kangra (32N/76E) earthquake occurred on 4 April 1905 and measured at least M8 on the Richter scale. It had effects throughout the northwest Himalaya, including the Kullu district, and was felt throughout India (*Civil and Military Gazette* 1905). Kullu was among the mountain areas most affected, and the impact illustrates cascading and interactive effects in such a hazard. Also, in the context of globalization since the 1990s, the example illustrates the importance of external links as an element of resilience.

The main shock struck Kullu at about six a.m. on 4 April 1905, a time at which many people were still indoors and during a season when most livestock were still housed in the lower floors and surroundings of the houses. In Kullu alone, 827 people died, over 10,000 livestock perished, 17,058 houses were destroyed, and 16,208 houses were damaged (*Punjab District Gazetteer* 1918). No fatalities occurred among local European settlers.

The area was relatively inaccessible: news of the devastation did not reach the "outside world" until 12 April, and the colonial administrator, Mr Calvert, who was away when the quake struck, was not able to reach Sultanpur, the capital and site of major damage, until 21 April, despite attempting a speedy return, because of the blockage of trails and tracks by landslides and snow avalanches. This was before the time of motorized transport. Damaging aftershocks continued well into July, and landslides continued to obstruct access and pose a hazard until the end of the monsoon season in September, as a result of the weakened sediments on hill slopes (*Civil and Military Gazette* 1905). Throughout the aftermath, outbreaks of cholera and other infections were reported in settlements in the area.

The example illustrates cascading effects initiated by the earthquake and its aftershocks, largely through landslides and other secondary and tertiary hazards, making even limited access more difficult. The emergency response was left entirely to the local community, as was the initial part of recovery and reconstruction. The reconstruction of destroyed houses was not completed for several years in some cases, and by the community and the district administration. One administrative action that assisted in reconstruction was the lifting of timber-cutting restrictions so that local people could replace destroyed and damaged structures. Evidence of built-in resilience is found in the fewer fatalities among people and livestock housed in the traditional wood and stone structures, which withstood the initial shock, as opposed to the newer masonry buildings in Sultanpur, which collapsed (*Punjab District Gazetteer* 1918). The impact on life and livelihoods in Kullu was enormous and felt for years, but the communities did "bounce back" and recover, largely through their own devices.

Many of the same qualities characterize earthquake disasters in the mountains in more recent times (Hewitt 1976; Slaymaker 2010), including secondary and tertiary hazards, differential structural damage and death, disruptions to access and retarded emergency response, and perhaps limited external support in long-term recovery for at least some sectors of the affected population. Significant advances in emergency response and recovery, and therefore resilience, have been made as a result of the rapid dissemination of information, better overland access, access by air, improved health care and public health, the increased strategic importance of many mountain areas, and new international organizations and agencies focused on disaster relief. For the most part, elements of new resilience in mountain social-ecological systems in the face of earthquake hazard result from external links and influences.

Yet the earthquake of 8 October 2005 in northern Pakistan shows again that some of the vulnerabilities present in 1905 remain and may be inherent to mountain regions. Mountain weather and terrain, continuing aftershocks and fear, landslides, broader institutional unpreparedness, lack of financial and other resources, and political sensitivities in the region slowed and hampered emergency relief and recovery. Large numbers of people remained without shelter long after the event, food stores were destroyed, and cold winter weather had secondary effects on the health and safety of the populace. Increased populations, incessant conflict, and the widespread building of unsafe structures, in contrast to traditional construction, in the area have exacerbated the impact beyond that experienced in 1905.

The four lessons from this example are:

- Mountain terrain and weather make relief and rescue extremely difficult, even in present circumstances, causing mountain communities to be especially vulnerable and lacking in resilience in the short term.
- Traditional and indigenous construction technologies provide some measure of protection in the face of earthquake hazard.
- The impacts of earthquakes are felt regionally, so that assistance from nearby is often not available.
- Earthquakes cause a number of secondary and tertiary hazards at the local level in the mountains that exacerbate and magnify the impact and impede relief and recovery.

Snow in Carpenter Creek, British Columbia

Snow is both a hazard and a resource in high mountain regions. Snow and snow avalanches have long disrupted the movements of people and transportation of goods and services, caused death and injury to people and animals, and damaged property and infrastructure. Also, the lack of snow in the mountains contributes to water shortages, localized droughts, failed winter tourism, and increased forest fire hazard in following seasons. In other words, mountain social-ecological systems must adjust to the effects of too much as well as too little snow.

Transportation delays caused by avalanches can affect a regional and national economy, as in Canada, where the main rail and highway transport routes traverse avalanche hazard areas in British Columbia (Woods and Marsh 1983; Campbell et al. 2007). Perhaps more important, too much or too little snow in mountain watersheds during the winter may lead to water surpluses (floods) and water shortages (hydrological droughts) in rivers, affecting regions well outside the mountains. At present, snow avalanches continue to pose a hazard to BC transportation, as they do in snowy mountain terrain elsewhere. In addition, the growth of skiing in its several forms and back country snowmobiling has exposed many recreationists to the hazard, and this sector is most affected through fatalities and injury today in British Columbia.

This bittersweet story of snow in the mountains could be illustrated with many examples. The role of snow in the early history of BC commercial mining shows some key points (Gardner 1986). Carpenter Creek (50N/118), also in the Kootenays, opened to silver mining in 1892 with a "rush" of several thousand miners appearing within a few months. The town of Sandon, which grew to as many as 5,000 people, came into being within two years, and it was linked by cart roads and a railway to mainline routes to convey people, goods, and ore. This rapidly emerging social-ecological system had no other reason for existing than the mining of galena, the ore of silver, lead, and zinc. Many newcomers had little experience of living and working in snowy, mountain country. The steep, forested slopes of Carpenter Creek valley provided a ready supply of construction materials for town and mines as well as fuel for heating, processing, and transportation. Uncut forest areas were burned to more clearly expose ore-bearing rocks (Gardner 1986).

The Columbian (Selkirk) Mountains of British Columbia are known for their avalanche hazard. An elevated hazard at town and mine sites and along routes between them confronted the miners in Carpenter Creek valley. In heavy snow years, such as 1904, avalanches injured and killed people; destroyed or damaged buildings, mine portals, and water flumes; and delayed transportation of ore within and out of the area, causing significant economic losses. In light snow years, such as 1910, water shortages curtailed mining and ore concentration, and extensive forest fires because of the drought impeded the subsequent summer's operations.

The social-ecological system was extremely vulnerable, substantially because of snow. At the same time, it exhibited remarkable resilience in managing snow as a water and power resource, using it to transport ore from mine sites difficult of access in winter through "raw hiding" along snowy and ice-lined trails and chutes and by adjusting the timing of transportation of ore and goods, of movements of people, and of mining activities to mitigate hazards (Gardner 1986).

The Carpenter Creek mining community had its ups and downs through the years, but it persisted until 1929–30. It collapsed not because of difficulties relating to snow but because of the slump in silver prices – an external, global factor.

The four lessons here are:

- Social-ecological systems may exacerbate an existing hazard through uninformed environmental alteration – in this case, clearing the forest on steep slopes.
- Pursuit of a particular livelihood activity, in this case mining, can put systems at extreme risk.
- Remarkable resilience can be demonstrated in mitigating the hazard and taking advantage of the beneficial aspects of the material or process.
- Local resilience in dealing with hazards may be entirely negated by lack of livelihood resilience in the face of pervasive global factors, such as falling commodity prices and economic downturns.

BC Wildfires of 2003

June through September 2003 was one of the worst forest fire seasons ever in British Columbia. More than 2,500 fires spread over a vast area of the mountainous interior (Filmon 2004; Beck and Simpson 2007). They destroyed approximately 260,000 hectares of forest and more than 400 houses and businesses, forced 45,000 people from their homes, and caused over $700 million in damage. Forest fires are not a new phenomenon in the BC mountains, but those of 2003 caused more damage and disruption than any up to this time. Fires occurred prior to European settlement and throughout the 200 years of settlement in the BC interior. For example, the summer of 1910 experienced widespread, destructive fires throughout the southern interior, and some of these impinged on mining operations in the Carpenter Creek valley and surrounding areas. Other years of widespread fires followed on several years of below-normal precipitation, which produced progressive drying of living and dead flammable material – 2003 was the fourth successive year of drought (Filmon 2004).

Other causal factors, in both bio-geophysical conditions and social-ecological systems, are now shaping the province's wildfire hazard. For at least 70 years, forest management practices have emphasized fire prevention, suppression, and control, as has been the case throughout much of mountainous North America. In the 1970s, this approach began to change as people realized that periodic fire performs useful functions in forest ecosystems' health and maintenance. Fire prevention and suppression interfered with these processes and allowed the build-up of flammable material in forested areas, heightening the risk of fire and large, rapidly expanding wildfire outbreaks. In some areas, such as national parks, controlled burns and removal of flammable material have entered into forest management practices, in part to prevent the outbreak of high-magnitude wildfires. Over the same period, as fuel built up, people, settlements, and other infrastructure expanded into the forested areas, raising levels of risk and vulnerability. The two decades prior to 2003 saw increasing numbers of problematic "interface fires" (Buchan 2004; Filmon 2004), which impinge on human settlements. This increasing hazard and vulnerability have been recognized for some time, but until 2003 few communities and governments factored them into land regulation and management practices.

The 2003 fire season, in particular the Okanagan Mountain Park Fire, which burned 239 houses in the southern suburbs of Kelowna in August (Heritage BC Stops 2010), may have changed that. Pictures of high-priced homes burning found their way quickly into the national and international media, drawing attention to what was really a localized disaster with relatively minor effects on life, property, and livelihoods. The visual aspects of the fire were spectacular and increased people's attention. Within days, regional, provincial, national, and even international links and partnerships came into play to assist in the emergency response. A year later, most of the destroyed and damaged buildings had been replaced and repaired through federal and provincial disaster subsidies and private insurance payments. The construction trades and materials suppliers thrived. It has been a remarkable demonstration of resilience in the face of disaster, coming about through external links. At the same time, flammable material has been removed, reducing risk levels for the near future.

However, many similar interfaces exist throughout mountainous North America. While one would hope that lessons from 2003 will inform practices there and elsewhere, new realities, such as widespread expansion of pine bark beetles, which kill trees and add to the dry fuel load, and a trend to warmer and drier weather, increase the hazard in general.

Three lessons from this example are:

- Wildfire hazard has been exacerbated by forest management practices of the recent past, and the vulnerability of human

settlements has increased as urban-fringe residences expand into forested areas.
- The spectacular visual effects of wildfire impinging on up-scale urban-fringe areas quickly get attention through the media, especially television, and this brings into play rapid responses, ranging from local to international.
- Visibility, cross-scale institutional links, effective management, and access to a variety of financial resources can ameliorate the effects of disaster in a remarkably short period, demonstrating high levels of resilience despite extreme vulnerability.

CONCLUSIONS

Mountain regions are subject to a variety of hazards and provide examples of many disasters. In part, this is a consequence of the bio-geophysical characteristics of the environment (Figure 5.1). Also, population increases and changes in the type and intensity of land use have transformed mountain social-ecological systems and exposed residents and property to risks from natural hazards, increasing vulnerability. These changes at the sub-regional and local levels often arise from larger-scale global factors and trends such as climate change, economic, institutional, and cultural globalization, technological change, war/conflict/terrorism, and pandemics and epidemics.

The purpose of this chapter has been to examine the resilience of mountain social-ecological systems in the context of hazards. Some literature suggests that resilience is the corollary of vulnerability; in other words, as resilience increases, vulnerability decreases. This review suggests a variant in the sense that vulnerability describes a condition of exposure while resilience refers to processes that come into play during and following an event, which allow a social-ecological system to carry on, perhaps in an altered state. This in turn may change vulnerability, or it may not. Resilient systems have the ability to learn from experience and make adjustments, along with a number of other characteristics that we have articulated (Figure 5.1).

In conclusion, five general points emerge:

- "Traditional" mountain social-ecological systems achieved higher levels of resilience primarily through avoidance by management of location and seasonal variability, diversification of agricultural products, and the use of variable and diverse micro-ecosystems.

- This worked well for localized hazards such as landslides, floods, and debris torrents, but not for regional hazards such as earthquakes, droughts, epidemics, and fires, which produce widespread and lagged interactive and cascading effects.
- Global and regional factors (e.g., climate change, institutional change, war, epidemics) may act to increase or decrease the vulnerability and resilience of mountain systems.
- While vulnerabilities may have increased in mountain systems in the twentieth century, resilience has increased as well through the emergence and effectiveness of cross-scale and other links and partnerships with other regions and organizations operating at national and international scales, and resilience has been facilitated by an ability to move information, materials, services, and resources over great distances quickly.
- Disasters still occur in mountain regions and resemble past ones. Despite the evidence of increased resilience, rugged and complex terrain, variable and extreme weather, distance and isolation, social, political, and economic inequalities, and poverty, marginality, and powerlessness magnify the impact of, and prolong recovery from, disaster.

REFERENCES

Ali, I., and D. Butz. 2003. "The Shimshal governance model – a community conserved area: a sense of cultural identity, a way of life." *Policy Matters* 12: 111–20.

Bankoff, G., G. Frerks, and D. Hilhorst. 2007. *Mapping Vulnerability: Disasters, Development and People*. London: EarthScan: 236.

Barnard, P.L., L.A. Owen, M.C. Sharma, and R.C. Finkel. 2001. "Natural and human-induced landsliding in the Garhwal Himalaya of northern India." *Geomorphology* 40: 21–35.

Barrows, H.H. 1923. "Geography as human ecology." *Annals of the Association of American Geographers* 13: 1–21.

Batterbury, S., and T. Forsyth. 1999. "Fighting back: human adaptations in marginal environments." *Environment*: 6–30.

Beck, J., and B. Simpson. 2007. "Wildfire threat analysis and the development of a fuel management strategy for British Columbia." *Proceedings of the 4th International Wildland Fire Conference*. Seville, Spain. www. fire.uni-freiburg.de/sevilla-2007/contributions/html/in/pais.html TS3

Berkes, F. 2007. "Understanding uncertainty and reducing vulnerability: lessons from resilience thinking." *Natural Hazards* 41 (2): 283–95.

Berkes, F., J. Colding, and C. Folke, eds. 2003. *Navigating Social-Ecological Systems: Building Resilience for Complexity and Change.* Cambridge: Cambridge University Press.

Bingeman, K., F. Berkes, and J.S. Gardner. 2004. "Institutional responses to development pressures: resilience of social-ecological systems in Himachal Pradesh, India." *International Journal for Sustainable Development and World Ecology* 11: 99–115.

Brooke, T.R. 2001. "The evolution of human adaptability paradigms: toward a biology of poverty." In A.H. Goodman and T.L. Leatherman, eds., *Building a New Biocultural Synthesis: Political-Economic Perspectives on Human Biology.* Ann Arbor: University of Michigan Press: 43–73.

Buchan, R. 2004. "Interface-fire hazard protection – friend or foe of sustainability?" In L. Taylor and A. Ryall, eds., *Sustainable Mountain Communities.* Banff: Banff Centre: 86–8.

Buckle, P., G. Mars, and S. Smale. 2000. "New approaches to assessing vulnerability and resilience." *Australian Journal of Emergency Management* winter: 8–14.

Burton, I., R.W. Kates, and G. White. 1978. *The Environment as Hazard.* New York: Oxford University Press.

Campbell, C., L. Bakermans, B. Jamieson, and C. Stetham. 2007. Public Safety Canada, Canadian Avalanche Centre. *Current and Future Snow Avalanche Threats and Mitigation Measures in Canada.* Calgary: University of Calgary.

Chapin, F.S., G. Peterson, F. Berkes, T.V. Callaghan, P. Angelstam, M. Apps, C. Beier, Y. Bergeron, A.S. Crepin, K. Danell, T. Elmqvist, C. Folke, B. Forbes, N. Fresco, G. Juday, J. Niemela, A. Shvidenko, and G. Whiteman. 2004. "Resilience and vulnerability of northern regions to social and environmental change." *Ambio* 33: 344–9.

Chapin, F.S., S. Trainor, S. Huntington, O. Lovecraft, A. Zavaleta, E. Natcher, D. McGuire, D. Nelson, J. Ray, L. Calef, M. Fresco, N. Huntington, H. Rupp, T. DeWilde, and L. Naylor. 2008. "Increasing wildfire in Alaska's boreal forest: pathways to potential solutions of a wicked problem." *BioScience* 58 (6): 531–40.

Civil and Military Gazette. 1894. Newspaper, Lahore, India.

– 1905. Newspaper, Lahore, India.

Colding, J., T. Elmqvist, and P. Olsson. 2003. "Living with disturbance: building resilience in social-ecological systems." In Berkes, Colding, and Folke (2003): 163–85.

Cox, T.E. 2000. "The intended and unintended consequences of AIDS prevention among Badi in Tulispur." *Himalayan Research Bulletin* 20: 23–30.

Davidson-Hunt, I., and F. Berkes. 2003. "Nature and society through the lens of resilience: toward a human-in-ecosystem perspective." In Berkes, Colding, and Folke (2003): 53–82.

de Scally, F., and J.S. Gardner. 1994. "Characteristics and mitigation of the snow avalanche hazard in Kaghan Valley, Pakistan Himalaya." *Natural Hazards* 9: 197–213.

Dekens, J. 2005. "Livelihood Change and Resilience Building: A Village Study from the Darjeeling Hills, Eastern Himalaya, India." MNRM thesis, Natural Resources Institute, University of Manitoba, Winnipeg.

Duffield, C., J.S. Gardner, F. Berkes, and R.B. Singh. 1998. "Local knowledge in the assessment of resource sustainability: case studies in Himachal Pradesh, India." *Mountain Research and Development* 18: 35–49.

Filmon, G. 2004. *Firestorm 2003: A Provincial Review.* Victoria: Government of British Columbia.

Firth, R. 1969. *Social Change in Tikopia: Re-study of a Polynesian Community after a Generation.* London: George Allan and Unwin.

Focus. 2009. *Disaster Risk Reduction: A Gender and Livelihood Perspective.* InfoResources Focus No. 2109, Aug.

Folke, C., J. Colding, and F. Berkes. 2003. "Synthesis: building resilience and adaptive capacity in social-ecological systems." In Berkes, Colding, and Folke (2003): 352–87.

Gardner, J.S. 1986. "Snow as a resource and hazard in early-twentieth-century mining, Selkirk Mountains, British Columbia." *Canadian Geographer* 30: 217–28.

– 2002. "Natural hazards risk in the Kullu District, Himachal Pradesh, India." *Geographical Review* 92: 282–306.

Gardner, J.S., and E. Saczuk. 2004. "Systems for hazards identification in high mountain areas: an example from the Kullu District, western Himalaya." *Journal of Mountain Science* 1: 115–27.

– J. Sinclair, F. Berkes, and R.B. Singh. 2002. "Accelerated tourism development and its impacts in Kullu-Manali, H.P., India." *Tourism Recreation Research* 27: 9–20.

Goodman, A.H., and T.L. Leatherman. 2001. "Traversing the chasm between biology and culture: an introduction." In A.H. Goodman and T.L. Leatherman, eds., *Building a New Biocultural Synthesis: Political-Economic Perspectives on Human Biology.* Ann Arbor: University of Michigan Press: 3–41.

Gordon, K. 2004. *The Slocan: Portrait of a Valley.* Winlaw, BC: Sono Nis Press.

Haddow, G., J. Bullock, and D. Coppola. 2008. *Introduction to Emergency Management.* 3rd ed. Boston: Butterworth-Heinemann.

Harcourt, A.F.P. 1871. *The Himalayan Districts of Kooloo, Lahoul and Spiti*. New Delhi: Vivek Publishing.

Harris, C. 1997. "Industry and the good life around Idaho Peak." In *The Resettlement of British Columbia: Essays on Colonialism and Geographical Change*. Vancouver: UBC Press: 194–218.

Heritage BC Stops. 2010. Okanagan Mountain Fire 2003. www. heritagebcstops.com/okanagan-tour/okanagan-mountain-fire-of-2003 (accessed 29 Nov. 2010).

Hewitt, K. 1976. "Earthquake hazards in the mountains." *Natural History* 85: 30–7.

– 1983. "The idea of calamity in a technocratic age." In K. Hewitt, ed., *Interpretations of Calamity: From the Viewpoint of Human Ecology*. London: Allen and Unwin: 3–32.

– 1984. "Ecotonal settlement and natural hazards in mountain regions: the case of earthquake risks." *Mountain Research and Development* 4: 31–7.

– 1997a. *Regions of Risk: A Geographical Introduction to Disasters*. London: Longman.

– 1997b. "Risks and disasters in mountain lands." In B. Messerli and J.D. Ives, eds., *Mountains of the World: A Global Priority*. New York: Parthenon Publishing: 371–408.

Hewitt, K., and I. Burton. 1971. *The Hazardousness of a Place: A Regional Ecology of Damaging Events*. Department of Geography Research Publication No. 6. Toronto: University of Toronto Press.

Holling, C.S. 1986. "Resilience of ecosystems: local surprise and global change." In W.C. Clark and R.E. Munn, eds., *Sustainable Development and the Biosphere*. New York: Cambridge University Press: 292–317.

Joshi, N., and V. Singh. 2010. *Traditional Ecological Knowledge of Mountain People: Foundation for Sustainable Development in the Hindu Kush–Himalayan Region*. Delhi: Daya Publishing House.

Kasperson, R.E., and J.X. Kasperson, eds. 2001. *Global Environmental Risk*. Tokyo/New York: UN University Press.

Klinenberg, S. 2002. *Heat Wave*. Chicago: University of Chicago Press.

Mann, J.M., ed. 1992. *AIDS in the World*. Cambridge, MA: Harvard University Press.

Nicholson, M.H. 1963. *Mountain Gloom and Mountain Glory*. New York: Norton.

Norris, F., et al. 2008. "Community resilience as a metaphor, theory, set of capabilities and disaster readiness." *American Journal of Community Psychology* 41 (1): 127–50.

Ostrom, E. 1992. *Crafting Institutions for Self-governing Irrigation Systems*. San Francisco: Institute for Contemporary Studies.

Pretty, J., and H. Ward. 2004. "Social capital and the environment." *World Development* 29: 209–27.

Punjab District Gazetteer. 1918. *Kangra District: Kullu and Saraj, Lahaul, Spiti – 1917*. Lahore: Superintendent of Government Printing.

Ramakrishnan, P.S., R. Boojh, K.G. Saxena, U.M. Chandrashekara, D. Depommier, S. Patnaik, O.P. Toky, A.K. Gangawar, and R. Gangwar. 2005. *One Sun, Two Worlds: An Ecological Journey*. New Delhi: UNESCO and Oxford & IBH.

Resilience Alliance. 2003. *Research on Resilience in Social-Ecological Systems: A Basis for Sustainability*. resalliance.org

Rizvi, J. 1999. *Trans-Himalayan Caravans: Merchant Princes and Peasant Traders in Ladakh*. New Delhi and New York: Oxford University Press.

Robledo, C., M. Fischler, and A. Patino. 2004. "Increasing the resilience of hillside communities in Bolivia: has vulnerability to climate change been reduced as a result of previous sustainable development cooperation?" *Mountain Research and Development* 24: 14–18.

Saberwal, V., and A. Chattre. 2001. "The parvati and the tragopan: conservation and development in the Great Himalayan National Park." *Himalayan Research Bulletin* 21: 79–88.

Seddon, D. 1995a. "AIDS in Nepal: issues for consideration." *Himalayan Research Bulletin* 15: 2–12.

– 1995b. "Migration between Nepal and India." In R. Cohen, ed., *The Cambridge Survey of World Migration*. Cambridge: Cambridge University Press: 367–70.

Slaymaker, O. 2010. "Mountain hazards." In I. Alcàntara-Ayala and A. Goudie, eds., *Geomorphological Hazards and Disaster Prevention*. Cambridge: Cambridge University Press: 33–47.

Steinberg, T. 2000. *Acts of God: The Unnatural History of Natural Disasters in America*. New York: Oxford University Press.

Tucker, R.P. 1982. "The forests of the western Himalaya: the legacy of the British colonial administration." *Journal of Forest History* 26: 112–23.

Turnbull, C. 1972. *The Mountain People*. New York: Simon and Schuster.

Wasson, R.J., N. Juyal, M. Jaiswal, M. McCulloch, M.M. Sarin, V. Jain, et al. 2008. "The mountain–lowland debate: deforestation and sediment transport in the upper Ganga catchment." *Journal of Environmental Management* 88 (1): 53–61.

Weichselgartner, J., and J. Sendzimir. 2004. "Resolving the paradox: food for thought on the wider dimensions of natural disasters." *Mountain Research and Development* 24: 4–9.

Wisner, B., P. Blaikie, T. Cannon, and I. Davis. 2004. *At Risk: Natural Hazards, People's Vulnerability, and Disasters*. 2nd ed. London: Routledge.

Woods, J.G., and J.S. Marsh. 1983. *Snow War: An Illustrated History of Rogers Pass, Glacier National Park, B.C.* Toronto: National and Provincial Parks Association of Canada.

6

Grassroots Participation versus Dictated Partnership: Anatomy of the Turkish Risk Management Reality

P. GÜLKAN AND A.N. KARANCI

INTRODUCTION: FEATURES OF DISASTER MANAGEMENT IN TURKEY

Even in the period following a realignment of its fundamental premise, the disaster management system in Turkey remains highly centralized and hierarchical and works according to a parent law supported by a wide array of regulations that are in the process of revision. The system has been criticized for being passively reactive because its administrative provisions relate almost entirely to post-disaster response. The underlying assumption would seem to be that other legal or administrative instruments were adequate for mitigation and social preparedness – on occasion a clearly incorrect hypothesis. Figure 6.1 outlines the organizational structure of Turkish disaster management. In this chapter we describe the management tiers of the figure.

Responsibility for response runs from bottom to top, from district to province and to the national level, depending on the scale of the event. The pattern reflects the statist tradition that has proven to have strong staying power. Small-scale disasters are handled first at the district level. The central district of a province is also a district but is administered by the governor of that city. If the disaster surpasses local capacity, the provincial governor, who is ex-officio head of the Provincial Disaster and Emergency Management Directorate, becomes involved in response and recovery (level 1 in Figure 6.1).

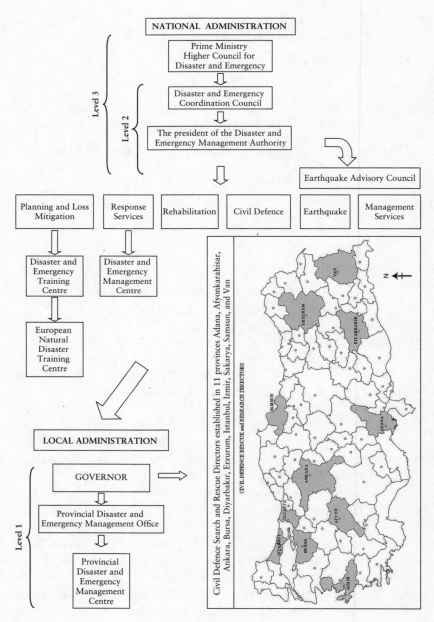

Figure 6.1
Outline of the National Disaster Management System, Turkey

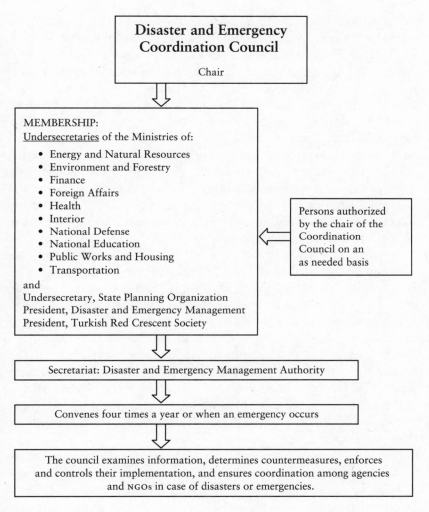

Figure 6.2
Disaster and Emergency Coordination Council, Turkey

When a major event requires national intervention, then the (Central) Disaster and Emergency Coordination Council, consisting of undersecretaries or high-level representatives from various ministries, coordinates response (see Figure 6.2). These measures were installed in principle first in 1959, long before large-scale urbanization and industrialization.

The national Coordinating Council consists of undersecretaries of related ministries and includes a representative from the Turkish General Staff and the president of the Red Crescent Society. If the prime minister decides that the size of the disaster necessitates a more comprehensive approach, he declares a state of crisis management; the Higher Council for Disaster and Emergency Management (level 3) – see Figure 6.3 – becomes effective, and a crisis centre is established in the prime minister's office. A regional crisis centre may be created if this seems necessary. The General Directorate of Disaster Affairs (GDDA) had been set up in 1959 within the Ministry of Reconstruction and Rehabilitation. The Emergency Management General Directorate (TAY) was created in 1999 in response to criticisms about fragmentation, especially at the local level, from international financing agencies.

A panicked government set up TAY after the two powerful and destructive earthquakes in 1999 but gave it little power. Because its responsibility overlapped with GDDA's, the new TAY played a minor role. The General Directorate of Civil Defence (GDCD) played a purely reactive role in search and rescue and in training community members or, recently, non-governmental organizations (NGOs) in disaster response. The government set up a national Earthquake Council in 2004 to unify these three agencies under the prime minister's office, and this became reality in December 2009, when GDDA, GDCD, and TAY fused as the Disaster and Emergency Management Authority of Turkey. Its six divisions shown in Figure 6.1 realign these agencies.

Provincial governors supervise the organizational structure for disaster management at their level. In response to a natural or human-caused disaster, each activates its Provincial Disaster and Emergency Management Office. There are several service groups within this body to implement diverse response and recovery efforts. Districts also establish the same structure. Some of these middle-level responsibilities are preparation and implementation of disaster response plans and implementation of training and exercising. Each level has its own financial resources. Thus municipalities and governorships at the provincial level have important responsibilities for mitigation:

- enforcement of earthquake-resistant design and other standards and regulations relating to land use planning for urban development

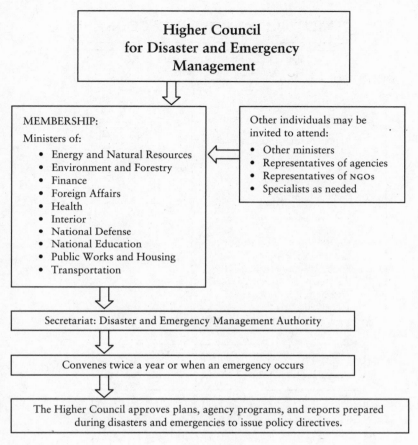

Figure 6.3
Higher Council for Disaster and Emergency Management, Turkey

- supervision of building construction
- post-disaster search and rescue

Effective from December 2009, local governments moved to the forefront of disaster management, but many of these are under-staffed and under-financed. Conspicuously absent are municipalities, which have the most effective instruments for disaster mitigation. For the Earthquake Advisory Council (central), see Figure 6.4; it has no municipal representation.

The next section examines the root causes of why disaster losses fall beyond the limits of societal acceptability of disaster losses.

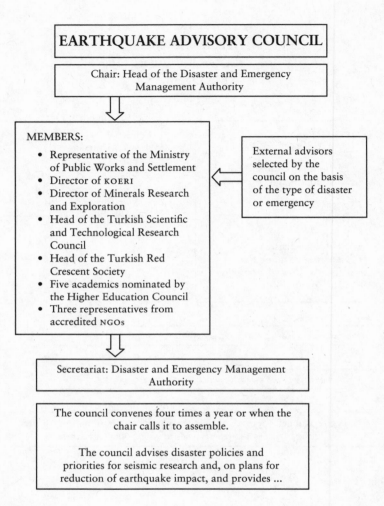

EARTHQUAKE ADVISORY COUNCIL

Chair: Head of the Disaster and Emergency
Management Authority

MEMBERS:

- Representative of the Ministry
  of Public Works and Settlement
- Director of KOERI
- Director of Minerals Research
  and Exploration
- Head of the Turkish Scientific
  and Technological Research
  Council
- Head of the Turkish Red
  Crescent Society
- Five academics nominated by
  the Higher Education Council
- Three representatives from
  accredited NGOs

External advisors
selected by the
council on the basis
of the type of disaster
or emergency

Secretariat: Disaster and Emergency Management
Authority

The council convenes four times a year or when the
chair calls it to assemble.

The council advises disaster policies and
priorities for seismic research and, on plans for
reduction of earthquake impact, and provides ...

Figure 6.4
Earthquake Advisory Council, Turkey

ROOT CAUSES FOR INADEQUATE PERFORMANCE

It is not clear whether new measures and developments since 1999 have proved effective. Recent small earthquakes, such as those in Elazığ in February 2010, and industrial and transportation accidents have continued to tax the capacity of the authorities, confirming continuing gaps in cooperation and coordination. There have been notable achievements in the number, quality, and distribution of professional and voluntary search and rescue teams, the national government has been considering replacing its policy of insuring all property at zero premiums with obligatory earthquake insurance, and legal reforms have aimed at creating effective supervision of building construction. There are more community organizations focusing on disaster preparation and response.

Unmet needs remain, however, and we summarize them below, under three headings: mitigation and preparedness, response and recovery, and national tasks.

Mitigation and Preparedness

Instead of containing damages caused by natural disasters through mitigation and preparedness, throughout the last few decades Turkey's government has emphasized response and recovery. It has not reduced losses in various sectors and has not met the heavy burden of each event because of insufficient resources. It is yet to create effective measures for distributing the financial burden through society. With the state ultimately footing all bills, public demand for loss reduction is small.

Structural retrofitting, essential in areas prone to earthquakes, has not been designed or initiated. Most people see building retrofit as reducing the resale value of property. There are no tax incentives, low interest rates, and so on to assist residents in this intervention. A few pilot programs in districts of Istanbul have attracted only a handful of volunteer property owners. Large-scale building surveys in Istanbul have left local governments unwilling to pay for improvements. Volunteer neighbourhood organizations exist for search and rescue, but none yet pushes for better construction standards or code enforcement.

Even though natural disasters have cost Turkey on average 1 per cent of gross national product (GNP) per year, the government has not yet linked development planning and disaster management. Consequently, it has neglected planning precautions to reduce the effects of disasters. Disasters corrode the economic base for rehabilitation, but economic planning has turned a deaf ear to this fact.

National research institutions have not done much to promote research and development to reduce losses from disasters. There has been little coordination of existing efforts, and the disparity between research and national needs remains.

National networks to monitor earthquakes and generate standardized data are rudimentary and inadequate, despite a country-
wide system for early damage estimation and for warning authorities
about a major earthquake. Responsible institutions do little to coop-
erate, coordinate efforts, or share data. As a consequence, Turkey has
no specific or multi-hazard techniques for assessing and analysing
vulnerability and risk. This has hampered local disaster planning.
The country possesses sufficient material resources and a reasonably
skilled workforce; it must organize, manage, plan, and mobilize these
resources to reduce disaster risk.

Rapid turnover in the Emergency Aid Service Groups in the
provincial committee may inhibit operations plans, skills, mock
exercises, and staff training at the local level. Despite significant improvements in disaster preparedness since 1999, search and rescue
teams lack personnel, rescue materials, and access to training. In
particular, first-response voluntary organizations and local people
are short of training and equipment, so probably would not be coordinated effectively in a disaster.

The hazards that Turkey faces in terms of earthquakes and other
disasters have not been adequately disseminated to the public. TV
channels entertain their audiences by inviting "experts" on the tenu-
ous connection between solar eclipses and earthquakes instead of
exploring the more mundane topics of safe urban planning or quality assurance in building construction. In universities, civil engineers,
architects, urban planners, and earth scientists have not been fully
trained in risks associated with the hazards of natural disasters and
on measures to reduce the risks. Existing facilities and capacities for
disaster management training have not been sufficiently mobilized
and coordinated. The increase in the number of learned articles has
caused illusory satisfaction when far more fundamental issues per-
sist: assuring building quality and enforcing codes.

Efforts to train government authorities, as well as managers and
planners, in disaster management, through continuous training and
exercises, have not been given the priority they deserve. Institutional
memory is fragile and unreliable. Although legislation and regulations in force date back a considerable time, disregarding them has
become commonplace, particularly in local government, through
violations of good building practices. The public is often a willing
partner in this process. There are no penalties or fines for violating
legislation and regulations. Public officials are in effect immune to

claims against their lapses of enforcement, and the legal system has no tools to protect the public from the consequences of negligence of duty or from the incompetence of professionals who provide services. Furthermore, due to lack of adequate awareness of civil rights, community members do not pursue their rights for full compensation.

Central and local governments have not been able to develop adequately planned settlement areas together with the proper infrastructure, leading to a gap between supply and demand in the construction sector. This gap has resulted in widespread illegal construction, with only nominal control or supervision of regulations by local governments. Illegal residential building has been indirectly encouraged by Parliament's unlawful construction amnesties. Construction legislation (about land use and building) has not been strictly followed in terms of physical planning and building construction inspection. This has aggravated the potential risks.

There are no rational balances between the duties of authorities and the responsibilities of the central government, and these duties have not been commensurate with the responsibilities of local government, the private sector, and the people involved in the prevention of natural disasters. Passing laws and regulations and establishing government agencies to reduce natural disasters are not sufficient. These agencies must receive the financial means to fulfil their mandates, and political will must fully endorse mitigation policies. Turkey is generally purely reactive vis-à-vis disasters, since public support and local sources are meagre for preparedness and mitigation. Healing the scars has served as the principal policy in response to disasters and has hindered community members' responsibility for and "ownership" of mitigation and preparedness.

The imbalance in the distribution of power and responsibilities is the most pervasive characteristic of the system, and certainly the most difficult to correct. While the system of disaster management requires the integrated cooperation and coordination of a number of ministries and other agencies, it does not possess instruments or mechanisms that would encourage and facilitate active participation and integration of the communities at risk in response and recovery. It is still highly paternalistic and gives assurances to people that the state will eventually replace all lost property, rebuild every shop, and rehabilitate affected investment through low-interest loans, debt annulment, and free credit. The concentration of population and economic activities in hazardous regions has increased the vulnerability of a larger segment of Turkish society in parallel to the progress in development that was evident when much of the current legislative fabric was formulated nearly 40 years ago. Social and economic progress has been increasingly affected by disasters, which have hindered growth. This situation increases vulnerability and threatens the environment, as has already happened in Kocaeli. Today, many cities face potentially destructive events of unprecedented dimensions. Hazard-conscious inter-sectoral planning is required to preserve their lifelines.

Response and Recovery

Provincial plans assume that local government employees who assist with rescue and relief are immune from the disaster. This assumption may disrupt effective response. As has been the case in many past earthquakes, members of local disaster management committees or their family members can be injured or otherwise incapacitated, lose their belongings or homes, and experience trauma when they could not help other people. Neither central nor provincial managers have the know-how to assess capacity and needs for effective and timely disaster response. Their positions are not permanent and often politicized. In some cases provincial officials are not from the province where they work and are unfamiliar with their surroundings. Rapid turnover of officials in some provinces may make response plans obsolete. Officials are often overworked and face other, more pressing priorities. Dispatching local public employees to disaster sites in neighbouring provinces may serve as a sensible and practical alternative. The hierarchical nature of disaster management tends to discourage local initiative.

   This situation raises another issue: minimal community participation and integration at all levels. Furthermore, the psycho-social needs of survivors and provision of appropriate support are largely uncoordinated, and there is a need to coordinate efforts of social workers, psychologists, and psychiatrists assessing needs with screening and psycho-social services. Following the 1999 earthquakes, six NGOs set up the Federation of Psycho-social Care in Disasters to help after disasters. The federation operates with the Turkish Red Crescent and has assisted in recent disasters, such as the Elazığ earthquake, floods, and mining accidents (Karanci 2009).

Despite the top-down structure of disaster management, coordination is lacking among government agencies, as became clear in the first few days following the earthquake of 17 August 1999. In addition, public organizations established many emergency operation centres during the first few days with no clear coordination and integration. Experience has shown inadequate central-provincial communication, coordination, and integration before, during, and after a natural disaster in terms of mitigation, preparedness, response, and recovery.

National Tasks

Sustainable development in such a disaster-prone country requires a sustainable mitigation strategy. The corresponding planning should involve all levels, particularly in physical planning. We are here proposing routes to an effective country-wide program.

The National Disaster Plan (Master Disaster Plan) prepared for the International Decade for Natural Disaster Reduction, or IDNDR (1990–9), must be revised and updated, particularly in the light of recent disasters, with suitable organizational structures and financial resources. An infrastructure of disaster management education would provide systematic and continuous training for central and local planners and managers. Training should be given to governors, deputy governors, district governors, mayors, and disaster-related civil bureaucrats. This training should be continuous and accompanied with drills or exercises.

Well-planned, sustainable programs of disaster training, with technical infrastructure, should start in primary and secondary schools. All relevant NGOs should participate.

Turkey should prioritize research and development at all phases of disaster management and coordinate these efforts. A sustainable strategy for risk reduction depends on active public participation. The strategy will have to focus on training, information dissemination, and raising public awareness, give training on specific skills for mitigation, and provide resources for community members to engage in actual mitigation behaviour. A risk culture and a resilient community need to be fostered. The fundamentals of Turkey's disaster hazards and risks and mitigation strategies should form part of the education system and be strictly followed, particularly for engineers, architects, planners, geologists, geophysicists, meteorologists, sociologists, psychologists, economists, and managers.

The psychological and sociological impact of disasters on communities deserves the highest consideration. A policy that addresses rehabilitation and reconstruction only vis-à-vis reconstruction or repair of damaged buildings cannot meet current needs and objectives. Therefore social, economic, and psychological rehabilitation of affected communities should be a goal, along with the creation of centres for social and psychological relief and support. Voluntary civic organizations such as those for youth and boy scouts can be encouraged to publicize disaster awareness so they can promote public awareness, community-based preparedness, and mitigation. This training should include members of the civil defence to ensure an adequate and skilled workforce in case of a severe disaster.

MECHANISMS FOR POPULAR INVOLVEMENT

Needs

Current disaster law and decrees do not bestow adequate responsibility on municipalities and NGOs. Effective risk management requires the involvement of all sectors (individual, community, local governments, public and private sector, and so on) and an all-hazards approach. For disaster preparedness, the governorship at the provincial level and the municipality alone are not sufficient. Individuals must prepare their own safety action plans and implement them. To this end, there must be community education on disaster risks, preparedness, fire services, search and rescue, first aid, disaster psychology, and networking (Consortium of Universities 2003). It is important to identify the target population (where people reside, socio-cultural characteristics, level of education, and so on) and to delineate channels and methods for reaching them.

The content of community education and social networking involves the following aspects:

- Communicate awareness of disasters to all segments of the community.
- Train and educate people about mitigation, preparedness, response, and recovery in order to enhance their capacity in disaster management.
- Conduct networking to strengthen connections between NGOs and to facilitate their participation in disaster management throughout all phases of the disaster cycle.

Community Education and Networking

In Turkey, awareness education and skills training for disasters (first aid, search and rescue, safe living in temporary accommodation after disaster, psychosocial support for survivors) were scarce prior to the Kocaeli and Düzce earthquakes of 1999. It was assumed that public institutions dealt with activities relating to preparedness, mitigation, and response. However, following 1999 it became apparent that the public sector could not cope and that local participation was necessary. Community education and networking have expanded since then. The Disaster and Emergency Management Authority provides disaster preparedness and search and rescue education to the community. Every year, selected community members are designated as civil defence volunteers and trained. These designated helpers seem to lack motivation to learn the necessary skills. Education and community participation and networking efforts have been under way since the 1990s but have had rather limited impact.

For example, the Earthquake Preparedness Pilot Project in Bursa in 1998 aimed to increase mitigation and preparedness in a province in the country's highest seismic hazard zone – one of the shaded regions in Figure 6.1 – by enhancing community awareness and participation. The program aimed to involve all relevant sectors. A working group consisted of representatives of local public and private sectors, municipalities, and NGOs. Local Agenda 21 staff, under the Bursa metropolitan municipality, facilitated the working group, which prepared an action plan.

To launch a community awareness program, a handbook for training trainers and a booklet for community members were prepared. However, there were some problems, especially in the relationship between the municipality and the local public sector, legally designated to handle disaster management. Also, there was little motivation to conduct mitigation and preparedness activities. Furthermore, local and general elections produced changes in the municipality, which led to the disbanding of the working group. When communities do not have adequate awareness and anxiety about future earthquakes, sustainable measures are impossible. Continuity of political and social will is essential to grassroots involvement.

Following the 1999 Kocaeli quakes, many NGOs organized activities within disaster-stricken areas. Some were involved in community disaster education and training, focusing on first aid and search and rescue. Thus there does not seem to be a systematic program that focuses on all phases of disaster management. The Turkish Psychological Society prepared booklets on psychological responses and coping strategies for different groups of survivors, which were distributed in the earthquake area to increase awareness by survivors. Some NGOs were established in the region to address the economic, social, psychological, and other difficulties of survivors.

In Kocaeli, a project to involve community members as volunteers has been initiated with the collaboration of the governor's office, the Provincial Directorate of Civil Defence, the İzmit municipality, Kocaeli University, and the Swiss Agency for International Cooperation. The Neighbourhood Disaster Support Project (MAG, after its acronym in Turkish) provided a very good example of local participation and planning for disaster management. District teams were formed, and networks between them were strengthened. The project aimed to train about 15 teams equipped with disaster kits and to form connections between the district support centres and the local crisis organization. The training included disaster awareness, disaster psychology, basic fire extinguishment, first aid, and search and rescue. Local institutions carried out the public education; for example, the Kocaeli Chamber of Physicians gave basic training in first aid, and the İzmit and Gölcük fire departments, in extinguishing fires. In Istanbul, as the community awareness component of ISMEP (Istanbul Seismic Risk Mitigation and Emergency Preparedness), Safe Life Volunteers training was launched in August 2009 to train community members in disaster mitigation and preparedness.[1]

As we can see in the above discussions, before the 1999 Marmara earthquake, Turkey had no systematic public education and training on disaster mitigation and preparedness. Trainers need extensive preparation, but there was a lack of standardized, culturally appropriate, and tested materials to prepare them. Materials need to consider different segments of the community and suit different levels. In order to reach all segments, the program needs to be institutionalized. The education and training programs organized by directorates of civil defence fail to reach all segments of the population, cover only response and recovery, and do not deal effectively with beliefs and societal actions on mitigation. Previous community education had a rather limited scope, but from these applications important principles can be derived. We have seen the possible role and impact of NGOs in community education following the 1999 earthquakes.

In Turkey, all municipalities, private business organizations, and NGOs can offer adult education on disaster mitigation and preparedness and run programs under the Ministry of Education. Since 1999, there have been major advances in community education and training, particularly in Istanbul, through the cooperation of the metropolitan municipality and district municipalities in collaboration with the NGOs. The institutionalization of these steps will render community education more systematic and extensive and more in keeping with scientific information. Education programs now deal with response and recovery. It is necessary to include mitigation, to coordinate agencies involved in community education, and to develop standards for programs. It is also essential to extend education to all segments of the population in urban settlements vulnerable to disasters. Another important issue is standards for training trainers and monitoring and evaluating educational activities.

CONCLUSIONS

The earthquakes in 1999 have now become the threshold events for comparisons between "old" and "new." The experiences have clearly laid bare the shortcomings and weaknesses of disaster management strategies and systems in Turkey.

A National Disaster Management Plan is needed. Although Law No. 7269 and associated legislation entrust the relevant organizations, provinces, and districts with preparing disaster plans, they do not explain either policies and strategies or the need for coordination and integration. In order to develop an effective disaster management system, Turkey needs to develop a basic National Disaster Management Strategic Plan where popular involvement is skilfully included. The country must "mainstream" mitigation policies in development planning and make the public an active stakeholder in this process. Work has been undertaken within the Disaster and Emergency Management Authority to develop this strategy.

National development and disaster management must be intertwined. Despite the fact that Turkey is a disaster-prone country, the link does not yet inform national development. The subject was included in the Eighth Five-Year Development Plan (2001–5), but not in the Ninth. These plans play an increasingly marginal role in the country's economic and social growth.

Compulsory earthquake insurance must be made law. The insurance that came into effect in 2000 covers only residential buildings within municipalities and covers only policy holders. Some areas require refinement: compulsory insurance should cover other common disasters such as flooding; there should be different premiums for buildings prone to different levels of risk so as to contribute to the building inspection system and encourage risk reduction. Compulsory insurance has been weakened by publicly funded construction programs for the homeless following earthquakes in Afyon in 2001 and Bingöl in 2003. Parliament passed into law in 2005 relief for affected rural houses in Karliova, a small town close to the triple junction of several active faults in eastern Turkey. Government assistance may be forthcoming in the future, even in the absence of insurance. In Elazığ in 2010, the M6 earthquake caused much destruction in a sparsely populated rural area, and the government undertook to construct cooperative-style buildings for homeless and uninsured citizens.

Participatory disaster management must be part of a social contract. Turkey has not formed the necessary mechanism for public participation and integration in disaster response. However, without active public participation and support, action for mitigation and preparedness is almost impossible. Effective relief and response require the support and active participation of the public and NGOs. There are no systematic and continuous public education and training on reducing disaster losses. Many initiatives, most of them individual efforts, started in 1999, but there is little coordination between them.

A major deficiency is the lack of accurate land use maps for evaluation of the local natural disaster hazard, so that more rational use could be planned. Local governments have tended to overlook this component when making decisions. Engineers and planners have long complained that urban vulnerability and environmental degradation have reached alarming proportions because policies on land use are absent. The State Planning Organization has remained indifferent to the threat of disaster.

Other major problems remain: the lack of supervision of building construction, from project stage to completion, and the disregard of legal responsibility for sub-standard building practices, including municipal approval of them without controls. Even since 2000, the Building Construction Law has covered only 19 "pilot" provinces. Beginning in 2011, it applies to the entire country, but there are many millions of inadequate buildings already in existence.

Earthquake-resistant design is not enforced in rural areas. Enforcement – for example, vis-à-vis materials standards – even in urban centres is controversial. Municipal building departments are not technically capable of providing final quality assurance for structural design. All the regulations imposed by the Development Law in the past decade produced an erroneous policy on the supervising system that encouraged only the technical application responsibility mechanism. In that system, the contractor paid the fee of the supervisor, who was theoretically expected to regulate his or her actions for the quality of the end product. After the 1999 earthquake, the government enacted first Decree No. 595 and then Law No. 4708 to change this policy. The regulations covered only 19 of the 81 provinces and excluded buildings up to two storeys in height and with less than 200 sq m in covered area. We hope that new amendments will help to solve these problems and meet actual needs, while expanding coverage to the entire country.[2]

NOTES

1 ISMEP is a World Bank–funded multi-year, pre-emptive program to make the metropolis better prepared to cope with a major earthquake likely to occur before 2030.
2 The Turkish Parliament has enacted substantial legal revisions since we prepared the initial draft of this paper. Those changes are not reflected in detail in this narrative.

REFERENCES

Consortium of Universities. 2003. *Earthquake Master Plan for Istanbul*. Istanbul: Metropolitan Municipality of Istanbul, July.
Karanci, A.N. 2009. "Psychosocial aspects of disaster management." In P. Gülkan and B. Başbuğ-Erkan, eds., *Perspectives in Disaster Management*. Ankara: METU Disaster Management Research Centre.

7

Disaster Management and Public Policies in Bangladesh: Institutional Partnerships in Cyclone Hazards Mitigation and Response

C. EMDAD HAQUE, MIZAN R. KHAN,
MOHAMMED SALIM UDDIN, AND
SAYEDUR R. CHOWDHURY

INTRODUCTION

Nature-triggered disasters, including extreme weather-related events such as heat waves, droughts, excessive precipitation, floods, and cyclones, have increased worldwide in recent years, both in frequency and in intensity. Such was the conclusion of the IPCC Assessment Reports (IPCC 2001a; 2001b; 2007b). Cyclones have intensified with increased sea surface temperature as a consequence of global warming (AMS 2006; WMO 2006; Emanuel 2008; Karim and Mimura 2008; Frank et al. 2010). Economic losses from disasters of all kinds are rising dramatically – almost nine-fold in real terms between 1960 and 1990 – and insured losses more than fifteen-fold. Of these, losses due to extreme precipitation, floods, and storms increased most (Munich Re Group 2003). Total US losses from Hurricane Katrina are expected to exceed US$125 billion, making it the world's most costly hurricane (ISDR 2005). The IPCC has also concluded that at least part of the increase in economic losses is the result of changes in climatic conditions (IPCC 2001b; 2007a). The other major factors are changes in land use, increasing concentration of people, and physical and infrastructural capital in vulnerable areas such as coastal regions, floodplains, and river basins (Miletti 1999; Nicholls et al. 2007; Solomon et al. 2009).

The geographical location of Bangladesh in South Asia, at the confluence of three large river systems – the Brahmaputra, the Ganges, and the Meghna – renders it one of the most vulnerable places to floods and cyclones. Human-induced climate change exacerbates the problem, with its already manifest effects and the predicted rise in sea level of 0.3 m to 0.5 m by 2050 (Agrawala et al. 2003; Government of Bangladesh 2005; Loucks et al. 2010). Climate models have revealed that the effects of climate change are not only affecting individual countries, but resulting in increased climate variations at regional levels (IPCC 2001a). Bangladesh, as part of South Asia, is likely to face more variations in climate regimes and thus extreme weather events.

In recent years, the intensity of drought has also increased in its northwest region, harming agriculture. Moreover, Bangladesh is the most densely populated country in the world, except city-states, with more than 1,000 people per sq km (Asian Development Bank and World Bank 2004). Agriculture, which provides a quarter of the gross domestic product (GDP), depends largely on nature. In the period 1970–2004, about 0.7 million people were killed, and economic losses totalled about US$5.5 billion (Chowdhury and Rahman 2001; Haque 2003; CRED 2004; FFWC 2005). The cyclone of 1970 alone, in the coastal areas of what was then East Pakistan, claimed over half a million lives. Again, the 1991 cyclone killed about 149,000 people.

In recent years, the frequency of extreme floods has increased, as well as the economic loss. The flood of 1998 inundated two-thirds of Bangladesh for two months. Dhaka City has become more vulnerable to floods. Over the years, the loss and damage to property have greatly increased, although human casualties have declined. Heavy rainfall has resulted in flash floods, which disrupt life by various means.

But on 11 June 2007, a heavy downpour triggering mudslides killed at least 106 people at the base of hills in Chittagong. The flood in July 2004 was the most devastating – with an economic loss of about US$2.2 billion (Asian Development Bank and World Bank 2004). This loss in terms of GDP was less than what the world's poorest countries faced during the 1985–99 disasters – a loss of 13.4 per cent of combined GDP (ISDR 2004). But the loss in Bangladesh amounted to a huge step back in development efforts. The floods in 2007 inundated about 36 per cent of the total area in 57 out of 64 districts (CEGIS 2007) and affected at least 4.5 million people (Oxfam 2007).

Because of extreme vulnerability to natural disasters, the government has developed institutions for disaster management (GOB 1997;

2004). These initiatives recognize the role of different stakeholders. For example, the Disaster Management Act, 1998, acknowledges the capacity of affected populations (GOB 1998): "An event, natural or man-made, sudden or progressive, that seriously disrupts the functioning of a society, causing ... such severity that the affected community has to respond by taking exceptional measures and is on a scale that exceeds the ability of the affected people to cope with using only its own resources."

Historically, non-government and community-based organizations (NGOs/CBOs) and other informal support mechanisms also have helped with disaster recovery. But disaster management warrants more than relief and recovery: it should be part of development planning, without which the loss of life and property is likely to intensify. Partnerships can be effective when they involve all stakeholders – government, local communities, NGO/CBOs, media, the private sector, academe, neighbouring countries, and donor communities.

Partnership in the present study represents a shift from a managerial approach to participatory, collective decision-making and resource sharing to manage disaster risk. Since communities are the direct and most seriously affected victims, effective and sustainable partnership requires a change from equal partnership to a focus on the community (Bhatt 2005). Thus partnership is a higher form of participation. Our central concerns relate to who decides, at what level, and how. How do people realize this approach effectively in practice? These are issues of a policy-institutional process.

There are numerous reports and action plans on natural disasters in Bangladesh. These materials deal mainly with the causes of disasters and make policy recommendations. Although the latter talk of an enhanced role for stakeholders, there is no analysis yet of how institutions at different levels are functioning. Is this mechanism based on partnership, with collective decision-making? Is a culture of working together on a national cause such as disaster management evolving? How functional are these elaborate institutional mechanisms? What is the role of the private sector or academe? How can an effective partnership be built into disaster management? These are questions the present chapter examines.

The method section is followed by a brief presentation of the institutional structure. The remaining two sections focus on analysis and a proposed framework of partnership.

METHOD

Our research aimed at analysing whether the elaborate institutional mechanisms of disaster management in Bangladesh reflect the partnership of stakeholders. For the purpose, social science policy analysis seemed useful. Obtaining reliable quantitative data on activities of both government and NGOs is a chronic problem in Bangladesh, so this study focuses on qualitative, rather than quantitative, parameters. The authors have taken the role of observers and interpreters.

Accordingly, content analysis has been our main method of research. The time frame was the years from 1970 to 2007. Within this period, Bangladesh experienced the most devastating cyclones in its recent history and several widespread, long-lasting floods that caused immense suffering to people and damage to properties. Official documents from the government and donors, study reports from NGOs and other organizations, journal articles, newspaper clippings, TV reports and documentaries, and internet resources from reliable and responsible sources provided information for our analysis.

To ensure openness in discussing sensitive issues, we used *informal* discussions with stakeholders at different levels. Through personal contacts and over the internet we collected reports and documents from government agencies, NGOs, and donors in Bangladesh, such as the Bangladesh Disaster Preparedness Centre (BDPC), the Disaster Forum, the Disaster Management Bureau, the Sustainable Development Resource Centre, and the United Nations Development Programme (UNDP). We obtained information from secondary and primary sources on:

- the extent and trend of disasters in Bangladesh and beyond
- the existing policy-institutional framework of disaster management
- the projects/programs of the government for disaster management
- the role and activities of various stakeholders
- the process of disaster policy-making and the functioning of institutional mechanisms

As we mentioned above, most of the information was qualitative in nature. Based on our analysis of the data and information, we propose a partnership framework and identify and spell out the roles of stakeholders.

INSTITUTIONAL FRAMEWORK

To manage the consequences of natural disasters, formal institutions in Bangladesh have formulated a well-developed mechanism (Figure 7.1) at national and local levels. Three factors led to such a development:

- The severity of casualties motivated local, national, and international levels to address the issue.
- Recurrent disasters set back development: loss in production and infrastructure hurt affected regions and the country.
- So that Bangladesh could attract external investment, minimization of disaster risks and vulnerabilities warranted intervention at the policy level.

Here is a brief description of the roles of major actors in the existing institutional mechanism. At the national level, there are four high-profile bodies for multisectoral coordination: the National Disaster Management Council (NDMC), headed by the prime minister; the Inter-Ministerial Disaster Management Coordination Committee (IMDMCC), led by the minister of food and disaster management; the National Disaster Management Advisory Committee (NDMAC), headed by a specialist nominated by the prime minister; and a Parliamentary Standing Committee on Disaster Management to supervise national policies and programs. The common missions of these bodies are to provide policy and management guidance and macro-coordination of activities, particularly vis-à-vis relief and rehabilitation.

The lead actor in disaster management is the Ministry of Food and Disaster Management (MOFDM), formerly the Ministry of Disaster Management and Relief (MODMR) until 2002. It has the role of interministerial planning of disaster management and coordination and of response in the event of a disaster. Under the MOFDM, there are two line agencies, the Disaster Management Bureau (DMB) and the Directorate of Relief and Rehabilitation (DRR). DMB is a small professional unit at the national level that performs specialist functions, working with district and *upazila* (sub-district) administrations and line ministries under the overall guidance of the IMDMCC. It is a catalyst for planning, for arranging public education, and for organizing systematic training of government officers and other personnel from the national down to the union (local council)/community level. The

DRR manages post-disaster provision of relief and rehabilitation. At present, it leads risk reduction at the community level.

Of all the other ministries, the Ministry of Water Resources (MOWR) plays a vital role in flood management. It plans water resources, including cyclone protection, flood proofing, riverbank erosion control, and drought management, although mitigation of disaster remains beyond its mandate. The Flood Forecasting and Warning Centre (FFWC), under the Bangladesh Water Development Board (BWDB) of the MOWR, provides early warning to the agencies involved.

Among other organizations, the Bangladesh Red Crescent Society (BRCS) and various donor agencies have notable roles. The Cyclone Preparedness Program (CPP) was established in 1972 following the devastating cyclone of 1970, under an agreement between the BRCS and the government of Bangladesh, to prepare for cyclones in coastal areas. CPP, under the BRCS, has a joint management structure, with two committees: a seven-member policy committee headed by the minister of MOFDM and a fifteen-member implementation board led by the secretary of MOFDM. Now the CPP has about 33,120 trained volunteers, including 5,520 women (MOFDM 2005).

Besides, the government has a "standing order" for natural disasters (mainly for floods and cyclones), which was last updated in August 1999. The standing orders govern all ministries, divisions/departments, and government agencies during normal times, precautionary and warning stages, the disaster, and afterwards.

Meanwhile, the Draft National Water Management Plan also stresses effective non-structural measures to reduce the impact of floods and erosion. Thus its participatory planning focuses on sustaining people's livelihoods, as opposed to implementing structural measures against floods (such as dams, river embankments, and flood control and drainage projects) and projects to control riverbank erosion (such as building of hard points, canalization, and revetment).

At the field level (Figure 7.1), disaster management and related mechanisms start with district administrations in the country's 64 districts. The District Disaster Management Committee (DDMC) is chaired by the deputy commissioner, the chief civil administrator. Members include departmental officers and representatives of BRCS, CPP, NGOs, and women. Likewise, below the district level, there are the upazila, union, and village tiers. These local bodies include representatives from almost all relevant interest groups (Figure 7.1). Subsections below examine how these committees function.

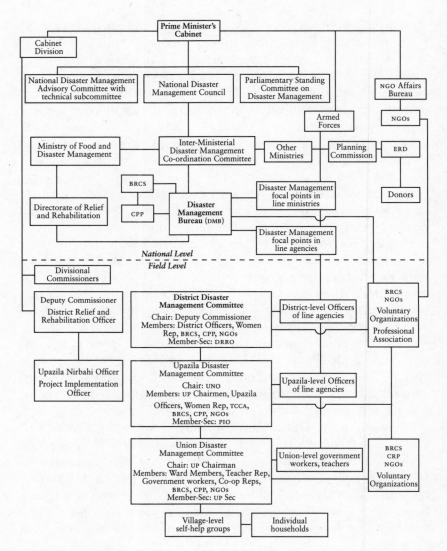

Figure 7.1
Disaster management at the national and field levels, Bangladesh: organizational structure and institutional arrangements

The Role of NGOs

The Disaster Management Bureau (DMB) coordinates the activities of NGOs – a vibrant sector in Bangladesh, acclaimed worldwide. NGOs and CBOS are active in, among other areas, disaster management, micro-credits, family planning, and human rights. NGOs in Bangladesh owe their origin to rehabilitation immediately after the war of independence of 1971. Currently, about a quarter of foreign assistance arrives through NGOs. Therefore their contribution, particularly to social services and mobilization of the poor, is quite prominent. NGOs such as the Grameen Bank (winner of the Nobel Peace Prize) and the Bangladesh Rural Advancement Committee (BRAC) have extended their programs at the international levels as well.

NGOs such as Action Aid, the Bangladesh Disaster Preparedness Centre (BDPC), CARE – Bangladesh, Disaster Forum, OXFAM – Bangladesh, and the Intermediate Technology Development Group – Bangladesh are particularly active in disaster management. Pre-disaster activities include advocacy, public education, and training for personnel in disaster management from the national down to the union or community level. NGOs also are active in emergency evacuation and in taking people to shelters. Post-disaster activities include offering new micro-credits or rescheduling loan payment programs for rehabilitation.

International Donor Agencies

International donor agencies are crucial in managing national and local development in Bangladesh, especially in the wake of disaster. International assistance arrives following each major disaster in response to the government's (GoB) requests for relief, rehabilitation, and reconstruction. Such assistance has been provided by UN agencies, intergovernmental bodies, and bilateral development partners. For example, following the flood of July 2004, the international community responded quite actively to relief and rehabilitation, despite the prevailing "aid fatigue." During the period from 17 July to 15 September 2004, some multilateral and many bilateral aid agencies provided disaster assistance to the tune of $76.3 million (UNDP 2005) – all figures here in US dollars. Following the ADB and World Bank Joint Assessment of the floods, ADB committed $180 million for rehabilitation. The World Bank as well pledged $200 million for infrastructure rehabilitation, livelihood restoration, and import financing (Asian Development Bank and World Bank 2004). During tropical cyclone SIDR of 15 November 2007, Bangladesh experienced roughly $1 billion in economic loss (MOFDM 2005; Daily Star 2007). Donor countries and agencies committed approximately $223 million for relief and rehabilitation (Moved 2005) to be disbursed through various channels, including the government and NGOs.

Surprisingly, more than six months after the cyclone in 2007, it was discovered that donor countries committing or pledging aid did not actually send their support. Only four countries of 32 sent part of this aid (Daily Ittefaq 2008). Afterwards, the attention of the world shifted to Myanmar, victim of another deadly cyclone, Nargis, on 2–3 May 2008, and days later the focus became China's earthquake in Sichuan province on 12 May. The non-fulfilment of commitments may represent a costly failure on the Bangladeshi government's part to maintain liaison with nations about their sending support. After more than six months, MOFDM and ERD of the MOFP were still just doing bureaucratic processing.

The donor community active in the development of Bangladesh has a consortium that listens to the government's development priorities and learns about Bangladesh's future strategic direction from the Bangladesh Development Forum (BDF). BDF is generally chaired by the minister/advisor of finance and planning and may also be co-chaired with a representative(s) from one or more major donor organizations. BDF is the top planning body in channelling aid to Bangladesh. The consortium – the Local Consultative Group (LCG) – acts as a forum. Thirty-two Bangladesh-based representatives of bilateral and multilateral donors of BDF compose this group, together with the secretary of the Economic Relations Division (ERD) of the Ministry of Finance and Planning (MOFP). The LCG prepares and follows up on BDF annual meetings and coordinates development in 21 sectors. One of these is Disaster and Emergency Response, which is looked after by the Disaster and Emergency Response (DER) sub-group.

DER acts as "a key national forum bringing together Government, NGOs, donors and UN agencies concerned with improving the effectiveness and efficiency of emergency response" (DER 2007). The UN resident coordinator in Bangladesh or a representative heads this sub-group. Since its formation in 2001, the World Food Program (WFP) has chaired this forum.

Moreover, there is a Disaster Response Group composed of donor representatives and headed by the UNDP resident representative in Dhaka. The development partners provide assistance to various aspects of the Flood Action Plan (FAP), which seeks to reduce the risk of severe flooding and enhance the technical capacity of the Storm Warning Centre of the Bangladesh Meteorological Department (BMD). Assistance from the International Federation of Red Cross/ Crescent Societies has helped sustain the Cyclone Preparedness Program (CPP) in disseminating cyclone warnings. Although local and national stakeholders have become more active in disaster management, the donors' role in funding the initiatives remains crucial.

The government's major initiative in the 1990s was the donor-aided Support to Disaster Management Project initiated in 1992 (DMB 1992). The DMB was created by this project and, since its inception in 1992, has enhanced national capacity to plan for, prepare for, and cope with the consequences of disasters. Its key achievements are:

- facilitating preparation of local disaster action plans (LDAPs) in 29 districts, 84 upazilas, 776 unions, and 24 *paurashavas/* municipalities
- forming and supporting disaster management committees (DMCs) at the union, upazila, and district levels
- drafting the Standing Orders on Disaster Preparedness, which define the functions of line ministries and departments at various stages of a disaster

The DMB has significantly improved efforts at saving lives and property from catastrophic floods and cyclones. Karim and Mimura (2008, 490) state that "over the last decade the country's capacity to deal with cyclones has improved considerably with the establishment of a cyclone warning and evacuation system. For example, cyclone SIDR in 2007 claimed far fewer lives (approx. 3,500) than the 1970 and 1991 cyclones, which killed at least 500,000 and 138,000 people, respectively."

Corporate Plan 2005–2009

Besides the CPP and the Standing Orders, the government adopted *Corporate Plan (2005–2009) – Comprehensive Disaster Management:*

Table 7.1
Community-Based Disaster Management Program (CDMP), Bangladesh

Strategic directions	Sub-programs	Target area/ group	Key outputs	Responsible agencies
Raising expertise of disaster management systems	Capacity building	MDMR and implementing agency	PPDU established and effectively executing its key functions New MDMR Allocation of Business and Organogram reflecting broader responsibilities in disaster risk management Professional skill enhancement program developed and implemented Professional training institutionalized Phase II program identified	MDMR/UNOPS
"Mainstreaming" programs for disaster risk management	Partnership development	National, district, and upazila officials	High-level advocacy program established and implemented Review of development project appraisal processes and integration of disaster risk management Training for national, district, and upazila officials implemented	MDMR/DMB/ NGO MDMR/DMB
Strengthening community institutional mechanisms	Community empowerment	Union, ward, and community levels	Inventory of existing programs developed and gaps identified Community risk management programs based on formal hazard analysis Local Risk Reduction Fund supporting community risk reduction	UNOPS DRR UNOPS
Expanding preparedness programs across a broad range of hazards	Research, information, and management	Dhaka and selected cities	Urban search and rescue pilot for Dhaka fire services based on earthquake threat Establishing an integrated approach to climate change risk management at national and local levels	MDMR/Fire Service MoEF/DoE
Operationalizing response systems	Response management	Whole country	Upgrading capacity in information management during normal and emergency periods Regional networks strengthened Timely deployment of resources Operational response capacities strengthened	MDMR

Source: Adapted from MoDMR (Ministry of Disaster Management and Relief) and UNDP. 2004. *Documents on Comprehensive Disaster Management Project.* Dhaka:

A Framework for Action (MOFDM 2005; see also MODMR and UNDP 2004). The $15–million Comprehensive Disaster Management Plan (CDMP, Table 7.1) is funded by the DFID and the UNDP. It aims "to reduce the level of community vulnerability to natural and human induced hazards and risks to manageable and humanitarian levels." This will be achieved through all aspects of risk management, in a transition from single-agency response and relief to a holistic strategy involving the entire development planning process of the government.

Corporate Plan 2005–2009 (MOFDM 2005) acknowledges the need for pre-disaster mitigation and preparedness as opposed to earlier concepts of post-disaster response. Priority has now been accorded to community-level preparedness, response, recovery, and rehabilitation through programs to train people living in disaster-prone areas and improve their capability to cope. *Corporate Plan* emphasizes three broad-based strategies.

First, disaster management would involve the *management of both risks and consequences* of disasters, which would include prevention, emergency response, and recovery. Second, *community involvement* in preparedness programs to protect lives and properties would be a major focus. The involvement of local government bodies would be essential. Self-reliance should be the key for preparedness, response, and recovery. Third, *non-structural mitigation measures*, such as community disaster preparedness, training, advocacy, and public awareness, must be given high priority; this would require the integration of structural mitigation with non-structural measures.

The CDMP lays the foundation for the shift from post-disaster relief and response to comprehensive risk minimization, which encourages disaster resilience through interconnected strategic directives:

- raising the level of expertise of the disaster management systems
- "mainstreaming" disaster risk management programming
- strengthening community institutional mechanisms
- expanding preparedness programs across a broad range of hazards, putting response systems into operation

The major sub-programs of CDMP include capacity-building, partnership development, community empowerment, research and information management, and response management.

Under the sub-program Partnership Development, the government would seek a multi-agency approach that would encompass the government, NGOs, and the private sector in a collaborative strategy to ease disaster-induced poverty. This would enhance coordination and information sharing and use resources effectively to reduce risk. Under the Community Empowerment sub-program, the government plans to expand the CDMP program and build community capacity through awareness and skill development and the expansion of disaster management studies within the school system and staff-training academies.

Disaster risk reduction has been incorporated into the Poverty Reduction Strategy Paper (PRSP) of Bangladesh as Annex-9 of Disaster Vulnerability and Risk Management (GOB 2004). The preparation of the PRSP, funded by the World Bank, acknowledges a holistic approach to disaster management.

ANALYSIS

Bangladesh appears to outsiders a land of natural disasters. Eighty per cent of its land is floodplains, and 92 per cent of its river flows, in terms of volume, originate from beyond its border. As a result, floods and cyclones have challenged this area since time immemorial. In fact, nature brings both blessings and curses – limited flooding keeps the soil fertile, or Bangladesh would not be able to feed such a huge population. Based on experiential learning, people withstand the adversity of nature, and this resilience is appreciated worldwide. Extended kin and family-based response to natural disasters are the basis of a strong culture. Together, all strata of society come forward and work in unity in relief and rehabilitation during and after natural disasters. Partnerships could build on this time-tested national attribute.

As is evident from its institutional structure, Bangladesh has developed an elaborate framework and disaster response mechanism. Moreover, some pronouncements in recent years indicate that the government is adopting an approach that factors in both risks and consequences. Disaster management capacities have improved during the last decades. After the 1988 floods and 1991 cyclone, disaster management replaced disaster control. The ministry became the Ministry of Disaster Management and Relief (MoDMR) in 1993 and then the Ministry of Food and Disaster Management (MoFDM) in 2002.

The DMB and the DRR lead in disaster management. The government's Rules of Business reflect the MoFDM's approach of comprehensive, community-based management to reduce vulnerability and risk. Despite declining losses in lives and property, particularly from cyclones, flood damage has tended to rise because of its wide area coverage, its increased frequency, and the expanding economy.

Government documents and the NGO literature indicate wide recognition that effective disaster response at the local level is not possible by government agencies alone and must involve all stakeholders. Still, the major lacuna in the institutional framework continues to be the lack of functioning partnership. The massive flood of July 2004 showed that no partnership or even coordination was functioning. The Local Consultative Group (LCG) found massive shortcomings in forecasting, preparedness, and coordinated response (LCG 2004). As a result, the NGOs conducted relief and rehabilitation largely without government directives and coordination. Initially, the government appeared confident to deal single-handedly with recovery. When things became worse, it made a flash appeal on 17 August 2004 through the UNDP, Dhaka, for international assistance. Another report about the 1998 floods indicates that "limited evidence of government coordination was found in the recovery phase" (World Bank 2005). Save the Children (USA) also noted "a general lack of coordination among actors" (Bacos et al. 1999). In the wake of cyclone SIDR in 2007, BBC (2007) reported, "Plenty of agencies, but not enough aid – Too little, too late," quoting a professional working in an affected area, "The reason why these people are not receiving enough help is because there is no coordination between the government and aid agencies." Similar failure has been observed in many other countries. For example, in the United States, intergovernmental planning and response failed to address shelter and housing requirements from Hurricane Katrina (Nigg, Bradshaw, and Torres 2006; Kapucu, Augustin, and Garayev 2009).

A striking example of poor management and coordination may be the following case. SIDR hit Southkhali village in Shoronkhola upazila of Bagerhat district in Bangladesh very hard. During a visit immediately after the event, the Indian foreign minister pledged his country's intention to rebuild all the houses in this and the surrounding villages. From then on, the government did nothing to give shelter to the affected people in this area and put virtually an official ban on others, including NGOs and aid agencies, from doing so. The Indian support did not come in, and even 100 days after the event people still had to live in the open (BBC 2008). This unfortunate situation arose perhaps from the lack of international/bilateral coordination, bureaucracy in both countries, lack of understanding of the victims' needs, and perhaps the government's unnecessary exercise of power.

The question therefore remains: why is there yet an absence of partnership? The following subsections seek to analyse some root causes to help find an answer.

Inadequate Policy

Bangladesh does not yet have any comprehensive policy for disaster management, despite bits and pieces in different programs and projects. As well, a partnership approach is lacking in the policy-institutional framework. This is the reason perhaps why the workshop organized by the Prime Minister's Office (PMO) after the flood of 2004 recommended "establishing effective GO-NGO coordination for ensuring better flood risk management before, during and after flood disaster" and "facilitating local governments and community involvement in flood preparedness activities through a participatory process" (PMO 2004).

The Strategy section of *Corporate Plan 2005–2009* (MoFDM 2005) makes no mention of partnership, but activities are expected to strengthen "Community Institutional mechanism (community empowerment)." *Corporate Plan* aims "to establish formal partnerships with government agencies, NGOs, civil society and the private sector for effective and sustainable service delivery." The government has not announced the kind of "formal partnerships" or how they would be implemented. In Bangladesh, the concerned ministry initiates the draft of a policy or act, places it for vetting in the Ministry of Law, and then forwards it to Parliament for debate and passage. In a few cases, policy formulation involves public hearings – for example, focus groups or focal point consultation, seminars and workshops, and public feedback through e-mail, fax, or letters. In the absence of Parliament, the president issued an executive order in the form of an ordinance.

However, Parliament has remained ineffective for most of the past 15 years thanks to harsh confrontations between the ruling and major opposition parties, which have resulted in walk-outs, boycotts, strikes, and protests on the streets. The result has been a continuous lack of healthy parliamentary debate on national issues. The CDMP, a donor-funded initiative, has good potential in policy/plan development, but its sustainability cannot be guaranteed without the government's involvement.

Capacity for policy-making is limited in most ministries and agencies, and so risk reduction and pre-disaster mitigation have been minimal in development projects and activities. Disaster mitigation remains ad hoc for the most part and not yet incorporated into regular project guidelines and operational procedures.

Weak DMCs

The national-level committees are comprised mainly of government representatives. Discussions with both government and community leaders indicate that the committees at different levels were supposed to include NGO/CBO representatives, but most are not yet formed, non-functional, or functional only in crises. No regular meeting of any committee from national to local level takes place except after a major disaster. Even committees with NGO/CBO representatives often change with a new government. The donor-aided project of 1992 initiated the formation of committees in some areas, but no follow-up occurred. The 30–member highest-level NDMC does not have any representative from NGO/CBOs, the private sector, academe, or the media (BDPC 2003). Only the secretary general of the BRCS remains a member of the IMDMCC. The composition of the NDMC and the IMDMCC reveals that the chair of the BRCS nominates two representatives in each committee. Except for the chair and the secretary general of the BRCS, all other members of these committees represent government ministries or agencies. While standing orders instruct almost all government agencies, they do not cover NGOs and CBOS (GOB 1997). This gap stands in the way of synergy and multiplier effects in disaster management. Nevertheless, the NGOs frequently arrange public education and training for personnel in disaster management from the national down to the union or community level.

Lack of Good Governance

In the bitter partisan politics in the country, there are few established norms and credible institutions in governance. Relief and rehabilitation therefore serve as a political instrument. Together, there is no documentation or public disclosure of information, particularly vis-à-vis internal resource mobilization and allocations for disaster management.

Individually or collectively through CBOs/NGOs, citizens actively participate in relief and rehabilitation, which stands in sharp contrast to the reported looting in New Orleans during and after Katrina (Rodriguez, Trainor, and Quarantelli 2006). Informal structures operate quite effectively in Bangladesh, while the lack of effective response by formal structures in Louisiana permitted the nightmare there.

Corporate bodies, banks, offices, and institutions donate generously to the Prime Minister's Relief Fund during and after disasters for relief and rehabilitation. However, these donations, received personally by the prime minister in full view of national TV, have some political overtones, and this practice is not transparent. The amounts raised and how those are distributed are never disclosed.

Rehabilitation projects, small or large, are contracted out to government politicians and party activists. Informal discussions revealed a widespread view that this fund was used to buy votes. Most government resources go to areas that support the regime, particularly those of political bigwigs. Opposition parliamentarians feel alienated. A written statement from a former member of Parliament observes: "A cell was opened in the Prime Minister's Secretariat to regulate all funds for infrastructural development works and for distribution of relief materials. In the name of PM's commitment or priority, those are distributed in a way to bribe people to vote for the ruling party in the next general election" (Quader 2006). A donor report also sees "corruption issues" as systemic vis-à-vis infrastructure projects in Bangladesh (DFID 2002).

Competition among Stakeholders

Some Islamic political parties, including Jamaat-e-Islami, participated in a coalition government (2001–6). They are at extreme odds with secular NGOs, including the Grameen Bank and BRAC, mainly because of two factors: these NGOs work for the alleviation of rural poverty, where women comprise over 85 per cent of their membership, and most of them seek Western aid. These Islamist groups oppose the empowerment of women.

Historically, NGOs' role in disaster management has always been significant. However, in recent years, the governance of NGOs and the alleged political bias of some have become high-profile issues in NGO-donor-government dialogue. The coalition government perceived that NGOs were largely supporters of the Awami League, and some probably were. In a recent interrogation in detention, the ex-general secretary and other top leaders of the Awami League, including a cousin of its leader, Sheikh Hasina, disclosed a ploy of engaging NGO workers on the street in favour of the group's political agenda. These workers belonged to one of the largest, most successful, and internationally reputed NGOs. These leaders were being interrogated by the Joint Interrogation Cell (JIC) under the current military-backed caretaker government and were earlier detained on various corruption charges. The audio tape of the interrogation intentionally or inadvertently surfaced on the internet (EnterBangla Blogs 2007).

Drawing on this example, we may conclude that at least some NGOs had links to political forces, including the Awami League, and that caused the Bangladesh Nationalist Party (BNP) and its coalition government to distrust NGOs. Some observers think NGO influence excessive, undermining the government's authority by fostering a parallel system of development and governance (Rahman 2006). Such divisive politics without well-established institutions and systems can be a major obstacle to a functioning partnership in disaster management.

Lack of Transparency in NGOs

Currently, over one-quarter of the foreign aid that Bangladesh receives – an average of about $1.5 billion per year – is channelled through the NGO sector. NGOs are held responsible by the national government for the country's poor state of governance. During informal discussions with us, some government officials and CBO leaders indicated that NGOs were no better in terms of accountability and transparency. Except for large, well-established international NGOs, such as Oxfam and the International Union for Conservation of Nature (IUCN), most are personality-based and do not foster group leadership. A recent study of NGOs by the World Bank, Dhaka, urges action in five areas: financial accountability and transparency; the role of the family of founder directors; the composition, tenure, and functions of boards; the quality of governance; and the strength of government departments monitoring NGOs (World Bank 2006).

Yet the release of foreign funds needs approval by the NGO Affairs Bureau in the Prime Minister's Office. The government tends to guide NGOs towards its political ends, rather than accepting them as partners in collective decision-making. This situation inhibits the forging of a real partnership. As a result, NGOs mainly deliver services and mobilize community groups and are not part of the policy-making process.

Limits of Local Governments

Many policy pronouncements, including *Corporate Plan* (MOFDM 2005), urge the strengthening of local governments (LGs). These bodies should have a very active and independent role in disaster management, because it is often LG representatives who first offer assistance. But LGs have no autonomy in decision-making or financial matters – the national government strictly controls them as a kind of "vote bank." This is why a new government always initiates LG elections. This dependence inhibits the building of partnerships, particularly at local levels, which is crucial for disaster management.

Absence of Organized Private Participation

As we mentioned above, the private sector contributes informally to relief during and after disasters but has no say in decision-making. It builds and conducts business in risk-prone areas, often ignoring government directives, and so incurs heavy economic losses in times of disaster. Only recently a recommendation surfaced to engage the private sector in "risk analysis and risk reduction studies" (PMO 2004).

Lack of Institutional Capacity, Expertise, and Memory

Government administrative cadres staff the MOFDM, DMB, and DRR, particularly at senior levels. People are selected in competitions and receive training on general administration at several phases of their career. The leaders have limited or no technical knowledge in disaster management. Further, these jobs are transferable at any time, and it is general practice to rotate personnel among ministries and agencies, so it is difficult to maintain institutional memory. There is limited technical expertise in the DMB and the DRR, so people there, lacking decision-making or managerial power, remain on the sidelines. This is also a serious problem in partnership building.

Why Talk about Participatory Management?

Looking at the institutional set-ups, which are not functional in most cases, we may well ask why the government talks about a participatory approach. The literature suggests four main reasons for its doing so (Thompson 1995): attempts by bureaucracies to survive; recognition of the failure of past approaches; successful participatory approaches by other organizations, including NGOs (Cernea 1991; Pretty and Chambers 1994; World Bank 1994); and pressure from the international aid community, since a top-down strategy has failed.

Aid dependence has been declining since the early 1990s – from about 5 per cent to 2 per cent of GDP – to about $1.5 billion a year. Donors support many programs in various aspects of flood management (World Bank 2005). The government's accountability therefore focuses towards donors rather than communities. Such aid dependence naturally encourages public agencies to follow donors' standards (Mitlin and Thompson 1995), rather than responding to NGO and communities' needs.

So the pronouncements and actions regarding partnership seem to result from pressures from within and without. NGO involvement in grassroots community programs, particularly in service delivery, brought better results in poverty alleviation and women's empowerment (World Bank 2005). Donor pressures reinforced these trends. An ethic or culture of collective decision-making has not yet taken root, so donor-driven initiatives for participatory management accomplish little that lasts.

CONCLUSIONS: PROPOSAL OF A PARTNERSHIP FRAMEWORK

The above discussion shows that the government talks about a comprehensive and integrated approach to disaster management. Response capacity has increased as a result. However, in the absence of stable and transparent institutions, this approach remains largely on paper. Individual stakeholders continue to make significant contributions, but synergy and multiplier effects are still missing. Our analysis shows no culture of partnership in disaster management. Divisive partisan politics and the lack of good governance prevent partnership among stakeholders. So this chapter proposes a partnership framework outlining new roles and responsibilities for major players. Implementation of the framework could lead to partnership in disaster management in Bangladesh.

A Framework Proposed

Disaster management is a nationwide affair, affecting every organization and citizen. The government lacks the resources for the wide scope of the tasks. Therefore a broad-based partnership involving all stakeholders is a desirable and realistic approach to realizing the full potential at all stages of disaster management – namely, prevention, preparedness, response, and recovery (Quarantelli 1990). Constituencies or interests include government ministries/agencies, Parliament and its Standing Committee on Disaster Management, NGOs/ CBOs, the private sector, the media, academe, donors, and neighbouring countries (Figure 7.2). The approach, involving multi-modal communication and interaction, would integrate the activities of stakeholders into a functional partnership.

Ministry of Food and Disaster Management (MoFDM)

This organization, directly supported by DMB and DRR, remains the pivot and would channel and coordinate all communications and activities between and among all the partners. This coordinated process, particularly during non-crisis periods, would shift the focus from post-disaster to pre-disaster risk management.

However, the MoFDM must ensure the transparency of its own and all other agencies developing the partnership. Public disclosure and documentation should be mandatory for all stakeholders, and MoFDM must publish reports regularly. The ultimate goal of the partnership is to enhance the investment and social capital for community empowerment against disaster risks.

MoFDM should:

· adopt a comprehensive national policy for disaster management, with clear guidelines for an effective partnership of all stakeholders
· coordinate the functions of disaster management and climate change communities, to help integrate prevention with preparedness, response, and recovery, in both short- and long-term perspectives (CDMP appears to be beginning in the right direction.)

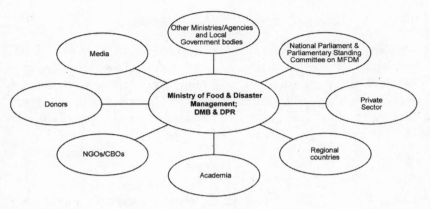

Figure 7.2
Partnership framework suggested for disaster management in Bangladesh

- strengthen the capacity of the MoFDM and other disaster management agencies and committees at all levels, particularly district-level disaster management committees (DDMCs). Each DDMC in risk-prone areas can be equipped with a geographical information system (GIS) cell as a planning tool for managing development and disaster reduction.
- train staff at all levels, including stakeholders, particularly media, NGOs, and the private sector, in team and motivational work and in how to prevent disasters; set up a Disaster Management Training Cell at the DMB
- activate DMCs at all levels, including national, through regular meetings
- strengthen project monitoring and evaluation at all levels, with the involvement of local stakeholders; establish broad-based and inclusive monitoring and evaluation committees for projects to ensure transparency, accountability, and delivery
- decentralize responsibilities and decision-making power to DDMCs, led by local governments with financial resources and autonomy
- establish small teams at all levels of DMCs to coordinate and integrate disaster management and develop a network among NGOs, GOs, researchers, academics, journalists, and other professionals to enlist their help in mitigating disaster-related problems, with MoFDM coordinating

NGOs/CBOs

This group has to be transparent and accountable to national and grassroots members. Once the areas of action and reform suggested by the World Bank study are taken into account, NGOs could exert pressure to translate government plans into action. Only then would they lead the process of involving local DMCs in all stages of disaster management.

Media

Media have published news, particularly on disasters' impact, but do little on risk reduction. The PMO recommends that they "highlight vital instruction in the media for people to follow" (PMO 2004). Media representatives need training on ways and means of reducing disaster risk so they can communicate effectively.

The Private Sector

Some builders and their associations realize they must factor earthquakes and land/mudslides into their work. The government has incorporated seismic risk into building codes but does not compel real estate developers to comply. Private investment, particularly in the risk-prone coastal belt, is vital, but government has said little on the subject. Builders, entrepreneurs, infrastructure developers, insurance agents, farmers, and tourist facility owners all should form part of decision-making and implementation through regular interaction.

In Sustainable Development and Disaster Risk Reduction for the poorest, the government promotes private activity in disaster management (SDRC 2004). Insurance is a standard mechanism for risk transfer, but access is difficult for the poor (ISDR 2004). This problem, along with weak social safety nets, results in a high level of vulnerability, which the risks of natural disaster exacerbate. Micro-insurance promises real potential for partnerships among public and private sectors at international, national, and local levels. Although micro-credit, pioneered in Bangladesh by the Grameen Bank, covers millions of women and poor people, it has not yet expanded to cover their assets through micro-insurance. NGOs' small offerings relate to life and credit insurance, with limited health coverage; however, the poor cannot build coverage from savings or micro-credits (Khan 2003). Communities, micro-finance institutions, donors, and private insurance and international reinsurance companies could set up pilot projects.

Academe

The relationship between the academic and research community and institutions for disaster reduction is just starting. Public and private universities could help with risk modelling, risk analysis, disaster recovery mechanisms, and training modules for capacity building. All stakeholders would benefit.

Parliament and Standing Committee

Parliament needs to be functional and less rabidly partisan. Disaster management is a national cause and requires bipartisan consensus over the basic parameters of policy-making and institution-building. The parliamentary committee must play its guiding and supervisory role. In this context, civil society, academe, and NGOs/CBOs need to press Parliament and its committees to become functional.

Regional Cooperation

Disaster management should be a function of regional cooperation involving Bhutan, China, India, and Nepal. Following the flood of July 2004, Bangladesh and Nepal formed a ten-member Joint Technical Study Team to work on flood mitigation and management. This team planned to draw up a work plan (*Daily Star* 2004) and submitted a report to the governments in 2006. Cooperation with India is vital, but its government has initiated a huge river-linking project to divert flow from the 38 trans-boundary rivers with sources in India, which might bring ecological disaster to Bangladesh (*Daily Star* 2005). At the ministerial level, Bangladesh has already expressed its concern to India. Under the circumstances, the region needs a framework of multilateral cooperation (PMO 2004), but India prefers bilateral deals. Bangladesh should promote diplomacy to establish a Regional Disaster Management Centre, which its prime minister has proposed.

REFERENCES

Agrawala, S., T. Ota, A.U. Ahmed, J. Smith, and M.V. Aalst. 2003. "Development and climate change in Bangladesh: focus on coastal flooding and the Sundarbans." Organization for Economic Co-operation and Development (OECD), Paris.

AMS (American Meteorological Society). 2006. "Is global climate change affecting hurricanes?" *AMS Video Journal*. www.ametsoc.org/videos. html

Asian Development Bank (ADB) and World Bank. 2004. *Bangladesh 2004 Post-Flood Recovery Programme: Damage and Needs Assessment.* Dhaka.

Bacos, D., M. O'Donnell, and M. Bennish. 1999. Disaster Response to the Bangladesh Floods of 1998. Dhaka: Save the Children. Mimeo.

BBC (British Broadcasting Corporation). 2007. "Plenty of agencies, but not enough aid." news.bbc.co.uk/2/hi/south_asia/ 7102020.stm (19 Nov.).

– 2008. "One hundred days after SIDR – very little progress in rehabilitation and housing" (an audio report in Bangla). www.bbc.co.uk/bengali/ indepth/story/2008/02/ 080222_mbsidr_100days.shtml (22 Feb.).

BDPC (Bangladesh Disaster Preparedness Centre). 2003. *Report on Problems of Coordination in Disaster Relief Operations in Bangladesh.* Presented to OXFAM, Dhaka.

Bhatt, M.R. 2004. "Community-based disaster risk reduction." In D. Etkin, ed., *Reducing Risk through Partnerships: Proceedings of the 1st CRHNet Symposium.* Winnipeg: 34.

CEGIS (Centre for Environmental and Geographic Information Services). 2007. "Bangladesh – flood affected areas." www.cegisbd.com/ flood2007/index.htm

Cernea, M. 1991. *Putting People First: Sociological Variables in Rural Development.* 2nd ed. New York: Oxford University Press for the World Bank.

Chowdhury, J.R., and R. Rahman. 2001. *Bangladesh Environment Outlook.* Dhaka.

CRED (Centre for Research on Epidemiology and Disasters). 2004. "An international disaster database." Brussels: Université catholique de Louvain. www.emdat.be

Daily Ittefaq. 2008. Dhaka. 21 May.

Daily Star. 2004. Dhaka. 31 Oct.

– 2005. Dhaka. 21 Oct.

– 2007. Dhaka. 25 Nov.

DER. 2007. "Disaster and emergency response (DER): a sub-group of LCG on emergency." www.lcgbangladesh.org/derweb/index.php

DFID (Department for International Development, UK). 2002. *Making Connections: Infrastructure for Poverty Reduction.* London.

DMB (Disaster Management Bureau). 1992. *Project Documents on Support to Disaster Management.* Dhaka: Government of Bangladesh.

Emanuel, K. 2008. "The hurricane-climate connection." *Bulletin of the American Meteorological Society* 89 (5): 10–20.

EnterBangla Blogs. 2007. enterbangla.blogspot.com/2007/06/corruption-in-bangladesh-and awami.html

FFWC (Flood Forecasting and Warning Centre). 2005. "An Overview of Flood Forecasting and Warning Services in Bangladesh." A paper presented on 2 April. Dhaka: Bangladesh Water Development Board.

Frank, D.C., J. Esper, C.C. Raible, U. Buntgen, V. Trouet, B. Stocker, and F. Joos. 2010. "Ensemble reconstruction constraints on the global carbon cycle sensitivity to climate." *Nature* 463 (7280): 527.

GOB (Government of Bangladesh). 1997. *Standing Orders on Disaster Management.* Dhaka: Disaster Management Bureau.

– 1998. *Disaster Management Act.* Dhaka: Disaster Management Bureau.

– 2004. *Poverty Reduction Strategy Paper (PRSP).* Dhaka: Ministry of Finance and Planning. Dec.

– 2005. *National Adaptation Program of Action.* Dhaka: Ministry of Environment and Forests.

Haque, C.E. 2003. "Perspectives of natural disasters in East and South Asia and the Pacific island states: socio-economic correlates and needs assessment." *Natural Hazards* 29: 465–83.

IPCC (Intergovernmental Panel on Climate Change). 2001a. In McCarthy et al., eds., *Climate Change 2001: Impacts, Adaptation and Vulnerability, Contribution of Working Group II to the Third Assessment Report.* Cambridge: Cambridge University Press.

– 2001b. In Houghton et al., eds., *Climate Change 2001: The Scientific Basis. Contribution of Working Group I to the IPCC Third Assessment Report.* Cambridge: Cambridge University Press.

– 2007a. *Climate Change 2007: Synthesis Report.* Geneva, Switzerland: Intergovernmental Panel on Climate Change: 104.

– 2007b. *Climate Change 2007: The Physical Science Basis.* Cambridge: Cambridge University Press: 996.

ISDR (International Strategy for Disaster Reduction). 2004. *Living with Risk: A Global Review of Disaster Reduction Initiatives.* Geneva. www.unisdr.org/eng/about_ isdr/bd-lwr-2004–eng.htm

– 2005. *Early Warning Newsletter*, issue 2005/3 (Sept)

Kapucu, N., M-E. Augustin, and V. Garayev. 2009. "Interstate partnerships in emergency management: emergency management assistance compact in response to catastrophic disasters." *Public Administration Review* 69(2): 297–313.

Karim, M.F., and N. Mimura. 2008. "Impacts of climate change and sea-level rise on cyclonic storm surge floods in Bangladesh." *Global Environmental Change* 18: 490–500.

Khan, M.R. 2003. Paper on Micro-insurance presented at the Workshop on Risk Assessment and Insurance, organized by the UNFCCC, Bonn, 12–15 May.

LCG (Local Consultative Group). 2004. "Notes on LCG environment meeting on lessons learned from floods of 1998 and 2004." Dhaka. 12 Aug.

Loucks, C., S. Barber-Meyer, M.A.A. Hossain, A. Barlow, and R.M. Chowdhury. 2010. "Sea level rise and tigers: predicted impacts to Bangladesh's Sundarbans mangroves a letter." *Climatic Change* 98: 291–8.

Miletti, D. 1999. *Disasters by Design*. Washington, DC: Joseph Henry Press.

Mitlin, D., and J. Thompson. 1995. "Participatory approaches in urban areas: strengthening civil society or reinforcing the status quo?" *Environment and Urbanization* 7(1): 231–50.

MODMR (Ministry of Disaster Management and Relief) and UNDP. 2004. *Documents on Comprehensive Disaster Management Project*. Dhaka: Government of Bangladesh.

MOFDM (Ministry of Food and Disaster Management). 2005. *Corporate Plan 2005–2009 – Comprehensive Disaster Management: A Framework for Action*. Dhaka: Government of Bangladesh.

Munich Re Group. 2003. *Natural Catastrophes in 2002*. Munich: Munich Re 10.

Nicholls, R.J., P.P. Wong, V.R. Burkett, J.O. Codignotto, J.E. Hay, R.F. McLean, S. Ragoonaden, and C.D. Woodroffe. 2007. "Coastal systems and low-lying areas." In M.L. Parry, O.F. Canziani, J.P. Palutikof, P.J. van der Linden, and C.E. Hanson, eds., *Climate Change 2007: Impacts, Adaptation and Vulnerability. Contribution of Working Group II to the Fourth Assessment Report of the Intergovernmental Panel on Climate Change*. Cambridge: Cambridge University Press: 315–56.

Nigg, J.M., J. Barnshaw, and M.R. Torres. 2006. "Hurricane Katrina and the flooding of New Orleans: emergent issues in sheltering and temporary housing." *Annals of the American Academy of Political and Social Science* 604(1): 113–28.

Oxfam. 2007. "South Asia floods, 2007." www.oxfam.org/en/programs/
emergencies/ southasia_floods_07/update_070806

PMO (Prime Minister's Office). 2004. *Options for Flood Risks and Damage Reduction in Bangladesh: Recommendations*. Dhaka: Government of Bangladesh. Sept.

Pretty, J.N., and R. Chambers. 1994. "Towards a learning paradigm: new professionalism and institutions for agriculture." In I. Scoones and J. Thompson, eds., *Beyond Farmer First: Rural Peoples' Knowledge, Agricultural Research and Extension Practice*. London: Intermediate Tech Publications Ltd.

Quader, G.M. 2006. "Democratic election?" *Daily Star*, 25 Sept.

Quarantelli, E.L. 1990. "Assessment of development potential and capacity based on vulnerability." In *Integrated Approach to Disaster Management and Regional Development: Planning with Peoples' Participation*. Dhaka: UN Centre for Regional Development.

Rahman, S. 2006. "Development, democracy and the NGO sector – theory and evidence from Bangladesh." *Journal of Developing Societies* 22 (4): 451–73.

Rodriguez, H., J. Trainor, and E.L. Quarantelli. 2006. "Rising to the challenges of a catastrophe: the emergent and prosocial behavior following Hurricane Katrina." *Annals of the American Academy of Political and Social Science* 604 (1): 82–101.

SDRC (Sustainable Development Resource Centre). 2004. Information provided to the authors regarding donor-supported projects on disaster management. Dhaka.

Solomon, S., G.-K. Plattner, R. Knutti, and P. Friedlingstein. 2009. "Irreversible climate change due to carbon dioxide emissions." *Proceedings of the National Academy of Sciences of the United States of America* 106 (6): 1704–9.

Thompson, J. 1995. "Participatory approaches in government bureaucracies: facilitating the process of institutional change." *World Development* 23 (9): 1521–54.

UNDP (United Nations Development Programme). 2005. *International Response to Flooding in Bangladesh*. Dhaka.

WMO (World Meteorological Organization). 2006. *Statement on Tropical Cyclones and Climate Change*. WMO International Workshop on Tropical Cyclones, IWTC-6, San Jose, Costa Rica, Nov.

World Bank. 1994. *A Strategy for Forest Sector in Sub-Saharan Africa*. Draft for discussion. Washington, DC.

– 2004. *World Development Indicators*. Washington, DC.

– 2005. T. Beck. *Learning Lessons from Disaster Recovery: The Case of Bangladesh*. Disaster Management Working Paper Series No. 11. Washington, DC.

– 2006. "Economy and governance of NGOs in Bangladesh." Dhaka.

PART THREE

Natural Hazards and Emergency Management in Canada

8

Emergency Management Education in Canada: A View from the Crossroads

LIANNE M. BELLISARIO, JACK McGEE, AND NIRU NIRUPAMA

INTRODUCTION

Consistent with global trends, the frequency of natural disasters in Canada is increasing (Etkin et al. 2004; International Federation of Red Cross and Red Crescent Societies 2004; Cutter and Emrich 2005). Urbanization, ageing and more diverse populations, the exploitation of vulnerable land, and reliance on increasingly interconnected and poorly maintained infrastructure are all contributing to increased societal vulnerability and more frequent disasters (Robert, Forget, and Rousselle 2003; Pielke and Sarewitz 2005). Add to this the expected impact of climate change on extreme weather events and wildfires (Houghton et al. 2001; Wotton, Martell, and Logan 2003; Nicholls et al. 2007), anticipated large-scale earthquakes (Clague, Bobrowsky, and Hyndman 1995; Hyndman 1995; Kanamori and Heaton 1996), and the growing threat of global terrorism, and it becomes clear that it is important to review the balance of risk in order to gain a deeper understanding of it. As a result of the greater grasp, identification, and evaluation of potential threats from natural, technological, and human-induced disasters, new plans for education, training, policies, and practices are emerging.

Further evidence of increasing risk can be seen as suburban areas press further into the countryside (Grossi and Kunreuther 2005). The risk of wild animals attacking residents and their pets increases (CBC Canada News 2005), stilt-type homes on hillsides become victim of landslides and earth tremors (Grossi and Windeler 2005; Krahn 2005), and communities built on floodplains continue to grow (Robert, Forget, and Rousselle 2003). Thus the complex inter- action among physical systems, the constructed environment, and the tendency of many people to live in attractive locations regardless of the risk or impact on ecological systems or the physical environ- ment suggests a need to address these concerns. In addition, Canada – especially Toronto and Vancouver – receives a continuous flow of new immigrants with professional degrees and experience in disaster management who do not obtain jobs in their respective fields. Therefore programs in disaster and emergency management, start- ing in high school, could generate public interest as well as demand for similar programs at university.

In the past decade, a greater range of risk factors has emerged, and their individual and combined effects have had a widespread impact both socially and economically. Canada has experienced bovine spon- giform encephalitis (BSE), severe acute respiratory syndrome (SARS), wildland–urban interface forest fires, Hurricane Juan, and the effects of the terrorist attacks in the United States in 2001. More private- and public-sector managers and specialists need to know how to prevent and manage the increasing variety, severity, and frequency of disasters.

More coverage of hazardous events by the media, growing aware- ness among the public, and globalization have increased perceptions of risk. As a consequence, Canada and the United States have made significant changes in government organization (Hoekstra 2003; Hwacha 2005). Greater emphasis on national security and public safety may lead to advances in emergency management, but only if the gap between practice and knowledge is addressed.

Evidence of this gap appeared in New Orleans after Hurricane Katrina. Reports indicated that, despite good documentation of the risk of disaster as a result of the levees being breached, little was done to prevent or prepare for such an event (Fischetti 2001; 2006). Michael T. Brown, former director of the Federal Emergency Management Agency (FEMA), told a US Senate hearing that, "after FEMA was folded into the Department of Homeland Security, there was a culture clash which didn't recognize the absolute, inherent sci- ence of preparing for disaster, responding to it, mitigating against future disasters and recovering from disasters. And any time that you break the cycle of preparing, responding, recovering and miti- gating, you are doomed to failure" (*New York Times* 2006).

In Canada, several universities and numerous community colleges have introduced programs in emergency management. However, further programs should involve practitioners, policy-makers, scientists, and law-makers. According to Ian Manock (2006), who teaches and administers the subject in an Australian university, "Today, as community and industry expectations of emergency managers grow, there is an increasing pressure on emergency management agencies to professionalize their staff through targeted tertiary education programs in their particular specialist fields."

It is in this context that we describe and evaluate emergency management education in Canada. The subject deals with risk and its avoidance (Haddow, Bullock, and Capploa 2008). This chapter builds on a dialogue that started at the first Canadian Risk and Hazards Network Symposium in Winnipeg, Manitoba, in November 2004. It explores current programs in Canada and discusses ways to develop a comprehensive, interdisciplinary, self-sustaining community in emergency management that is responsive to and reflective of Canadian experiences and values. Phillips (2003) calls emergency management "the process through which risk is managed in order to protect life and property through a comprehensive effort that involves nonlinear activities related to mitigation, preparedness, response and recovery." We look at the full spectrum of education opportunities, from training through to postgraduate research and teaching.

THE PAST: EMERGENCY MANAGEMENT EDUCATION IN CANADA

Emergency management was once the exclusive domain of military or quasi-military civil defence organizations, but civilians are now often responsible for it at the municipal, provincial, and federal levels. They must have the knowledge, skills, and experience to help mitigate, prepare for, respond to, and recover from a wide range of risks, including those stemming from natural hazards, climate change–induced events, biological threats and pandemics, technological hazards, and terrorist acts. In addition, the private sector needs employees with specialized knowledge and training in the area, even outside traditionally high-risk industries (Bruce et al. 2004). Both public and private sectors must plan for disasters. Private-sector employees should update their skills and must be prepared and resilient in case of any emergency (Wolff and Koenig 2010). Private-sector managers keen to minimize business interruption have broadly embraced emergency management principles as part of their planning for business continuity.

Despite this apparent widespread demand, emergency management has evolved more slowly in Canada than in Britain (Britton 2003; Stuart-Black, Norman, and Coles 2004), Australia (Manock 2006), and the United States (Neal 2000; Phillips 2003). In Canada, emergency management begins with a local response. When a community is unable to cope, it will call on its provincial or territorial government, which in turn may call on the federal government (Health Canada 2009). Canada has not regularly experienced disasters that have resulted in significant loss of life. In the United States, each hurricane season seems to bring more and more fatalities (the heat wave of 1980 and Hurricane Katrina were the two worst since 1900), and in Europe between 22,000 and 35,000 people died in the 2003 heat wave (International Federation of Red Cross and Red Crescent Societies 2004). In Australia, wild land–urban interface fires are a significant risk, as in those around Canberra in 2003 and in Victoria in 2009. Heat waves, floods, and tropical cyclones in that country have killed far more people than bushfires in the last 100 years. However, in overall impact on people, property, and the environment, drought tops the list of natural disasters there, while wildfires rank last (EM-DAT 2010).

In Canada, the relatively few fatalities spread out over many years have allowed memories to fade and have probably reinforced a traditional, reactive model of emergency management, which has done little to drive education or research. The slow development of the discipline may also be a result of public ignorance and indifference, although the success of programs at Brandon, Royal Roads, and York universities suggests a positive shift. Nevertheless, debate continues about "training"[1] versus "higher education,"[2] resulting in disagreement on how to spend scarce resources, the polarization of the small community of professionals, and barriers to cooperation. As Bruce et al. (2004) observe, this "us versus them" mentality has slowed program development.

Almost everyone in emergency management would agree that the historical lack of a strong educational infrastructure slowed the development of both the profession and the discipline in Canada (Bruce et al. 2004). Current programs suggest likely professionalization of the field. According to D.E. Alexander (2000), "although knowledge does not guarantee power over natural catastrophe, it is a prime requisite of disaster prevention." The Hyogo Framework stresses knowledge of and education on disaster risk reduction, which would lead to disaster-resilient communities and nations. Similarly, the United Nations International Strategy for Disaster Reduction (UNISDR) also urges the incorporation of disaster awareness and risk reduction in school curricula in vulnerable countries (Clerveaux and Spence 2009).

In Canada, even with the new programs, knowledge brokers – trainers, educators, and researchers – are in short supply, to say nothing of the near-complete absence of Canadian textbooks or shareable teaching resources (Bruce et al. 2004). Consequently, educational development has been slow and tended to focus on training first responders (e.g., Barg 2004). The Justice Institute of British Columbia (JIBC) was the first to develop certificate programs, in 2003. It noted that there were 250,000 front-line responders across Canada who needed postsecondary education in hazards and disaster/emergency management (McGee 2004).

In 2001, Brandon University began offering four-year degrees with a major in applied disaster and emergency studies (ADES) in 2001. The program aims to produce highly qualified professionals who can work in an interactive, dynamic, and high-pressure environment that demands solutions. The ADES degree was designed to combine social and physical science perspectives of hazards and disasters within a liberal arts education (Lindsay 2005a). It incorporates elements of modern society and the environment as they pertain to risk, disaster, emergency processes, and responses. It further stresses the principles of mitigation, organizational and resource planning, and management (Nirupama 2004).

The Stuart Nesbitt White Fellowship, set up in the 1960s by what is now Emergency Preparedness Canada, offers $13,500 annually to postgraduate students in a field relating to emergency management – for example, geography or sociology. Some former winners have themselves contributed to such programs (Bruce et al. 2004).

THE PRESENT: CHALLENGES AND OPPORTUNITIES

Research and recent experience now show that the traditional reactive approach to emergency management can go only so far in reducing risk. Since the United Nations International Decade for Natural Disaster Reduction (1990–9), the concept of mitigation as a key component of emergency management in Canada has been gaining ground (Hwacha 2005; Canadian Council of Forest Ministers Assistant Deputy Ministers Task Group 2006). Provinces are legislating hazard, risk, and vulnerability assessments and aiming to reduce risk proactively in programs, policies, and legislation – e.g., Québec's Loi sur la Sécurité civile of 2001; Ontario's Emergency Management Act of 2002; and British Columbia's Provincial Emergency Program of 2003.

These changes necessitate more comprehensive education in emergency management. In some countries, this might involve an undergraduate degree or training or certification by a nationally recognized body of professionals. In Canada, the path is much less clear. Universities and colleges are developing programs at various levels – certificate, undergraduate, and graduate. In Ontario, four colleges – Centennial, Fleming, George Brown, and Humber – all offer certificate programs. In the autumn of 2005, former Atkinson College at York University in Toronto enrolled its first students in a certificate program, which it developed in consultation with the Ontario Association of Emergency Managers (Ontario Association of Emergency Managers 2004). The Northern Alberta Institute of Technology, in consultation with a variety of national stakeholders, has developed a draft competence profile that describes required skills and training (such as government, health care, industry, military, security, or workplace health and safety) for emergency managers as a foundation for a two-year diploma program (Northern Alberta Institute of Technology 2005). Certificate programs are numerous, require less time, emphasize operational skills and knowledge, support themselves financially, and typically incorporate new technologies. However, without university research in the field, certificate programs will have to rely on other sources to build and renew their curricula.

Royal Roads University's executive-style master of arts in disaster and emergency management has been very successful. It uses an online platform to offer basic and advanced knowledge modules and courses, complemented by three weeks on campus for exercises and personal interaction. York University now offers a full-time master of arts in disaster and emergency management in the School of Administrative Studies. In 2010, York also launched a bachelor's program with specialized honours, honours, and general four-year and three-year degrees.

How to sustain and provide programming for emergency managers while building this field and creating a Canadian body of knowledge, expertise, and professionalism through research and degree programs? Achieving this balance of theory, research, and practice is critical for the sustainability and applicability of these academic programs.

Recruiting educators with Canadian expertise and knowledge remains a challenge. There are but a handful of potential candidates. Strong competition for personnel among institutions may slow the development of programs as new openings emerge. However, it may also draw in experienced new faculty members looking to develop programs and research. The shortage of educators and researchers, the small body of literature, and strong competition for scarce expertise and resources suggest the continuing slow development of the field in Canada. Some creative approaches are therefore needed to reinforce the collective effort of this evolving community.

THE FUTURE: BUILDING A COMMUNITY

Elected officials and the general public are expressing higher expectations of emergency management than ever before. Opportunities for a new generation of well-educated professionals will probably continue to grow as baby boomers retire, the demand for more comprehensive knowledge increases, and the need to counter the disaster mythology often perpetuated by the media grows (Fisher 1998; de Ville de Goyet 2000; Thevenot 2006).

Furthermore, given the pressures of climate change, urbanization, and increasing vulnerability and disaster costs, emergency management as a discipline should continue to evolve as Canadians search for new and innovative ways to reduce risk. Rather than competing for resources, training and higher education programs will ideally evolve together, or at least be connected, and considered legitimate components of a single whole. A necessary first step would be collective recognition that education encompasses both applied and theoretical components.

Programs in Canada appear to be well under way and gaining momentum, which must be sustained through more research and development. The Canadian Institute of Health Research (CIHR), the

Natural Sciences and Engineering Council of Canada (NSERC), and the Social Science and Humanities Council of Canada (SSHRC) encourage and fund research and partnerships in many new multidisciplinary studies. We in the field, as a collective body of academics, practitioners, policy- and decision-makers, members of NGOs, and other stakeholders at large, should continue to strengthen this fascinating field of study. Building Canadian knowledge, expertise, and experiential content is a key piece of the educational infrastructure. The fact that emergency management often embraces a body of knowledge and theories from a variety of other disciplines (Phillips 2004) works to the advantage of future education programs. The United States has graduated master's and doctorate-level students who have focused on emergency management within their home departments of geography, psychology, sociology, and urban planning (Neal 2000). These departments encouraged natural connections through discussions on sustainable development, natural hazards, risk perception, and social and organizational behaviour by allowing for supervision of graduate work in the field without approaching emergency management as a separate academic discipline.

Similar opportunities to leverage expertise in Canada are emerging and will soon be realized. In Canada, neither emergency management nor disaster studies is a distinct field of research for any of the academic funding councils (CIHR, NSERC, and SSHRC), as it is in the United States (e.g., the US National Science Foundation's Small Grants for Exploratory Research program). This lack of recognition translates into a limited pool of potential supervisors. Public Safety and Emergency Preparedness Canada recently expanded the Stewart Nesbitt White fellowship, which can now award up to eight postgraduate scholarships worth $19,250 annually. The Canadian Risk and Hazards Network (CRHNet) may help focus expertise and knowledge to the benefit of educational programs; it allows people in hazards research and education and in emergency management to share knowledge and innovative approaches that reduce vulnerability to disaster. It reaches out to both practitioners and researchers and may become an influential part of the emergency management community, working to mobilize resources and initiatives on its behalf and helping bridge the gap between theory and practice.

No discussion would be complete without comment on standards and certification. While leaving comprehensive examination to others (e.g., Alexander 2003), we note the absence of national standards for curriculum content in emergency management. Undergraduate programs follow commonly acceptable academic standards of Canadian universities. The Canadian Standards Association (CSA) has appropriately revised and announced new standards on emergency management by integrating it with business continuity planning (CSA 2010).

CONCLUDING REMARKS

It has been said that disasters represent a collective challenge that calls for a collective solution. Greater attention must be paid to identifying hazards, the vulnerabilities of people, and mitigation efforts to reduce risk and vulnerabilities, since responding after the fact is no longer sufficient. People working in emergency management need to understand not only the likely long-term impact of natural hazards, technological and human-induced hazards, and terrorist acts, but also why these events can cause a disaster in one area but a manageable emergency in another (Moseley 2004). They must know the theoretical concepts and ever-evolving human behaviour (including cultural sensitivity) and be able to apply their knowledge in a real-life setting. To ensure emergency management in Canada fits its milieu and to help create a culture of sustainable disaster mitigation, as promoted by Mileti (1999) and others, we must continue to educate people who should know the topic best and be most familiar with the Canadian disaster risk environment.

Human activities that cause emergency situations (e.g., poor land use planning) can be managed better if the essential knowledge, tools, and methods are understood and applied. Thus there is a significant role for education that focuses on disaster and emergency mitigation, preparedness, response, and recovery. We expect that all stakeholders ultimately hope to see the professionalization of emergency management in Canada, and strong training and academic programs are its prerequisites. The basic and advanced knowledge provided through educational programs will help professionals think from a different perspective, know ways to mitigate the dangers from any emergency, and appreciate the need for resilient communities.

While the number of researchers and instructors remains small, there is a growing realization that Canadian universities, colleges, and training institutions must collaborate and support development. We must also engage academicians in a discussion of how best to leverage their expertise to support this emerging discipline. There are many competing priorities for investments. Some observers might ask about the consequences if we do not work together to develop a solid education infrastructure in Canada. Quite simply, we risk being unable to modify the trend of increasing disaster impact and being caught sorely unprepared for the next event. Unless the risks associated with various human activities are understood and assessed in context by experts who have been taught to manage them, people will continue to suffer, and the loss of human lives and property will reinforce public distrust and the belief that nothing can be done. With an ever-increasing number of threats facing Canadians and with national security and public safety at the forefront of the political agenda, now more than ever we must ensure that there is a clear path for incorporating emergency management – theory, practice, and principles – into sound social, political, and economic decision-making at all levels. Supporting and strengthening the growing education community is a critical first step.

What can be done to enhance collaboration and accelerate the development of emergency management in Canada? Some possibilities are clear: an all-encompassing education program must incorporate teaching, research, and training; it must provide a foundation in theory, research, policy, and practice; and the education community, being small, will have to strike an alliance with all interested stakeholders. In our pursuit of a more robust education infrastructure, we can also learn from existing course materials and programs, as has been discovered in the United States (Neal 2000). Since the development of actual programs can be a difficult and long-term process (as we saw above), researchers and educators should establish informal cross-disciplinary specializations and research networks that build on existing strengths. Any new program needs acceptance by practitioners and academics and must also be viable within the academic bureaucracy. In order to develop a field of study and work, professionals in Canadian universities and colleges and practitioners must interact. Liaisons can bring to light field problems that need solutions and spur the creation of new knowledge (Neal 2000). These types of exchanges should be encouraged by employers (public and private sector), since a small investment – for example, allowing employees to advise students, teach, or participate in a research project – would allow their organization to be aware of and immediately benefit from advances in the discipline. Organizations may want to consider a mentoring program whereby practitioners would participate in local educational initiatives as part of their work responsibilities. This could go a long way towards ensuring continuity and communication between today's generation of emergency managers and tomorrow's.

Knowledge of emergency management (a risk management process) is necessary for a wide variety of careers today. The need for "hazard and disaster sensitive" professionals in other fields has been recognized (Blanchard 2003). Working emergency managers represent a substantial proportion of the students pursuing certificates and diplomas. The development of shareable course materials and resources, particularly for distance-learning students, whose requirements are often resource-intensive to develop – see Rubin (2003) and Neal (2004) – could be open to collaboration among institutions. The creation of an information "clearing-house" would also help leverage limited resources by providing a home (virtual or otherwise) for people seeking materials or wanting to share their own. There is an opportunity today to bring this community together to work towards a common long-term goal – the development of emergency management as a discipline and profession where members apply their expertise to reduce risk for all Canadians. Achieving that goal will require significant commitments by academe, governments, practitioners, and students. Much can be gained by simply creating vehicles through which members of the community can self-identify and interact. Bibliographies, discussion boards, listservs, seed money for research programs, and workshops could significantly enhance the educational infrastructure in Canada and solidify links among members. Universities with resources may want to take up and develop specializations in specific emergency management modules, resulting in less competition and greater capacity as a whole. This would allow for the leveraging of finite resources and expertise and foster a common approach. Overall, the development of mechanisms and incentives for encouraging these activities should be a priority for governments and private-sector entities that consider themselves members of this community of practice.

NOTES

1 The transfer of skills from trainer to learner (Kuban 2001).
2 Formal instruction leading to a degree or diploma at a recognized university or college.

REFERENCES

Alexander, D. 2003. "Towards the development of standards in emergency management training and education." *Disaster Prevention and Management* 12(2): 113–23.

Alexander, D.E. 2000. *Confronting Catastrophe: New Perspectives on Natural Disasters*. New York: Oxford University Press.

Barg, R. 2004. "Breaking down barriers: collaborative education drives collective change." *Journal of Emergency Management* 2 (3): 51–5.

Blanchard, W. 2003. "The new role of higher education in emergency management." *Journal of Emergency Management* 1 (2): 30–4.

Britton, N. 2003. "Higher education in emergency management: what is happening elsewhere?" www.training.fema.gov/EMIweb/downloads/ Neil%20Britton%20-%20Higher%20Education%20in%20Disaster %20Management1.doc

Bruce, J., K. Donovan, M. Hornof, and S. Barthos. 2004. "Emergency management education in Canada." www.psepc-sppcc.gc.ca/research/ resactivites/emerMan/2003D021_e.asp

Canadian Council of Forest Ministers Assistant Deputy Ministers Task Group. 2006. *Canadian Wildland Fire Strategy: A Vision for an Innovative and Integrated Approach to Managing the Risks*. Edmonton: Canadian Council of Forest Ministers, Natural Resources Canada, Canadian Forest Service.

CBC Canada News. 2005. "Alberta closes trails to keep people away from bears." 6 July. www.cbc.ca/story/Canada/national/2005/07/06/alta-Trail-050605.html

Clague, J.J., P.T. Bobrowsky, and R.D. Hyndman. 1995. "The threat of a great earthquake in southwestern British Columbia." *BC Professional Engineer* 46 (9): 4–8.

Clerveaux, V., and B. Spence. 2009. "The communication of disaster information and knowledge to children using game technique: the disaster awareness game." *International Journal of Environmental Research* 3 (2): 209–22.

CSA (Canadian Standards Association). 2010. www.csa.ca

Cutter, S.L., and C. Emrich. 2005. "Are natural hazards and disaster losses in the US increasing?" *EOS, Transactions, American Geophysical Union* 86 (41): 381–96.

De Ville de Goyet, C. 2000. "Stop propagating disaster myths." *Lancet* 356 (9231): 762.

EM-DAT. 2010. The International Disaster Database. www.emdat.be/result-country-profile

Etkin, D., E. Haque, L. Bellisario, and I. Burton. 2004. "An assessment of natural hazards and disasters in Canada: a report for decision-makers and practitioners." www.crhnet.ca/docs/Hazards_Assessment_Summary_eng.pdf

Fischetti, M. 2001. "Drowning New Orleans." *Scientific American* Oct.: 77–85.

– 2006. "Protecting New Orleans." *Scientific American* Feb.: 64–71.

Fisher, H.W. 1998. *Behavioral Response to Disaster: Fact versus Fiction and Its Perpetuation: The Sociology of Disaster.* 2nd ed. Landham, MD: University Press of America, Inc.

Grossi, P., and H. Kunreuther, eds. 2005. *Catastrophe Modeling: A New Approach to Managing Risk.* New York: Springer.

Grossi, P., and D. Windeler. 2005. "Sources, nature, and impact of uncertainties on catastrophe modeling." In Grossi and Kunreuther (2005): 69–92.

Haddow, G.D., J.A. Bullock, and D.P. Cappola. 2008. *Introduction to Emergency Management.* Oxford: Butterworth-Heinemann.

Health Canada. 2009. "Emergency management: taking a health perspective." *Health Policy Research Bulletin* 2009(15): 1–48

Henstra, D. 2003. "Federal emergency management in Canada and the United States after 11 September 2001." *Canadian Public Administration* 46(1): 103–16.

Houghton, J.T., Y. Ding, D.J. Griggs, M. Noguer, P.J. van der Linden, X. Dai, K. Maskell, and C.A. Johnson. 2001. *Climate Change 2001: The Scientific Basis.* Cambridge: Cambridge University Press.

Hwacha, V. 2005. "Canada's experience in developing a national disaster mitigation strategy: a deliberative dialogue approach." *Mitigation and Adaptation Strategies for Global Change* 10: 507–23.

Hyndman, R.D. 1995. "Giant earthquakes of the Pacific Northwest." *Scientific American* 273: 68–75.

International Federation of Red Cross and Red Crescent Societies. 2004. *World Disasters Report 2004: Focus on Community Resilience.* Bloomfield, CT: Kumarian Press, Inc.

Kanamori, H., and T.H. Heaton. 1996. "The wake of a legendary earthquake." *Nature* 379: 203–4.

Krahn, J. 2005. "Why do slopes become unstable after rainfall events?" *Geotechnical Fabrics Report* 23 (6): 20–1.

Kuban, R. 2001. "Dialogue on crisis: the need for 'education' too." www.
    iclr.org/pdf/research%20paper%2016%20-%20paper%205%
    20ron%20kuban.doc.pdf
Lindsay, J. 2005a. "Applied disaster and emergency studies: an overview of
    Brandon Universitys' ADES Department." Presentation at the Emergency
    Management Education in Canada Workshop, Toronto, 16 Nov.
–   2005b. "What's in a name?" Presentation at the Canadian Risk and
    Hazards Network Symposium, Toronto, 17–18 Nov.
Manock, I.D. 2006. "Tertiary emergency management education in
    Australia." training.fema.gov/EMIweb/edu/imanock.pdf
McGee, J. 2004. "The need for emergency management post secondary
    education." Paper at the Canadian Risk and Hazards Network
    Symposium, Winnipeg, 18 Nov.
Mileti, D.S. 1999. *Disasters by Design: A Reassessment of Natural
    Hazards in the United States*. Washington, DC: Joseph Henry Press.
Moseley, L. 2004. "Educational needs for disaster management." *Australian
    Journal of Emergency Management* 19 (4): 28–31.
Neal, D.M. 2000. "Developing degree programs in disaster management:
    some reflections and observations." *International Journal of Mass
    Emergencies and Disasters* 18 (3): 417–37.
–   2004. "Teaching introduction to disaster management: a comparison of
    classroom and virtual environments." *International Journal of Mass
    Emergencies and Disasters* 22 (1): 103–16.
New York Times. 2006. "Former FEMA chief's opening statement." 10 Feb.
    www.nytimes.com/2006/02/10/national/national special/10 Brown-
    statement.html
Nicholls, R.J., P.P. Wong, V.R. Burkett, J.O. Codignotto, J.E. Hay, R.F.
    McLean, S. Ragoonaden, and C.D. Woodroffe. 2007. "Coastal systems
    and low-lying areas." In M.L. Parry, O.F. Canziani, J.P. Palutikof, P.J.
    van der Linden, and C.E. Hanson, eds., *Climate Change 2007: Impacts,
    Adaptation and Vulnerability. Contribution of Working Group II to the
    Fourth Assessment Report of the Intergovernmental Panel on Climate
    Change*. Cambridge: Cambridge University Press: 315–56.
Nirupama, N. 2004. "Higher education issues in disaster studies." Paper at
    the Canadian Risk and Hazards Network Symposium, Winnipeg, 18 Nov.
Northern Alberta Institute of Technology (NAIT). 2005. *Draft Emergency
    Management Practitioner Competency Profile*. 5 April.
Ontario Association of Emergency Managers (OAEM). 2004. *Emergency
    Management Education Survey Report*. Aug.

Phillips, B. 2003. "Disasters by discipline: necessary dialogue for emergency management education." Paper at the Creating Educational Opportunities for the Hazard Manager of the 21st century workshop, Denver, CO, 22 Oct.

– 2004. "Using online tools to foster holistic participatory recovery: an educational approach." *Australian Journal of Emergency Management* 19 (4): 32–6.

Pielke, R.A., Jr, and D. Sarewitz. 2005. "Bringing society back into the climate debate." *Population and Environment* 26 (3): 255–68.

Robert, B., S. Forget, and J. Rousselle. 2003. "The effectiveness of flood damage reduction measures in the Montreal region." *Natural Hazards* 28: 367–85.

Rubin, C.B. 2003. "The need for digital educational resources in emergency management." *International Journal of Emergency Management* 1 (3): 309–16.

Stuart-Black, J., S. Norman, and E. Coles. 2004. "Bridging the divide from theory to practice." Paper at the 40th Anniversary Conference of the Disaster Research Center, University of Delaware, Newark, 30 April–1 May.

Thevenot, B. 2006. "Myth-making in New Orleans." *American Journalism Review* 27(6): 30–7.

Wolff, E., and G. Koenig. 2010. "The role of the private sector in emergency preparedness, planning and response." In E.B. Abbott and O.J. Hetzel, eds., *Homeland Security and Emergency Management: A Legal Guide for State and Local Governments.* Chicago: American Bar Association: 121–54.

Wotton, B.M., D.M. Martell, and K.A. Logan. 2003. "Climate change and people-caused forest fire occurrence in Ontario." *Climatic Change* 60: 275–95.

9

Public and Expert Knowledge and Perception of Climate Change–Induced Disaster Risk: Canadian Prairie Perspectives

PARNALI DHAR CHOWDHURY, C. EMDAD
HAQUE, AND GRAHAM SMITH

INTRODUCTION

This chapter proposes a theoretical and empirical framework for investigating the juxtaposition, overlaps, and gaps in public and experts' knowledge and perception of climate change–induced environmental extremes and their associated disaster risks. The empirical context involves three types of disaster risk relating to climate change or variability – floods, droughts, and heat waves – in the prairie region of Canada. The province of Manitoba occupies a large part of the region, and its location predisposes it to climate extremes and considerable climate change–induced disaster risks (Francis and Hengeveld 1998; IPCC 2007). The commonalities and gaps between residents and experts are relevant, because both groups make decisions about policies, programs, and implementation. Once the public is aware of the risks, it is likely to address the problem, including by demanding policy change (Weber 2006; Moser 2007).

Morgan et al. (2002) and Sterman and Sweeney (2007) investigated the knowledge and perception of graduate students, staff, and experts at Carnegie Mellon University (CMU), in Pittsburgh, Pennsylvania; laypeople (teenagers and parents); and the adult general population. Even relatively well-educated individuals conceptualized climate change very differently from experts. The public emphasized the depletion of the stratospheric ozone layer by chlorofluorocarbons (CFCs) and often disregarded the chief culprit – the build-up of carbon dioxide caused by burning fossil fuels. Morgan et al. compared the public conceptualization of key terms and definitions but did not compare it with that of the experts and revealed an implicit bias towards the latter. As Lidskog (2008) observes, our society depends on experts and their knowledge.

The research outcomes apply Cox et al.'s (2003) "generic methodology on the application of Mental Models" to a phenomenon where considerable uncertainty exists – climate change. The effectiveness of risk communication is relatively unknown, as well. The Cox approach builds on work at CMU and at Technische Universitat Berlin. The CMU model inclines asymmetrically towards the expert mental model and was initially an "influence diagram" reflecting experts' knowledge of relationships between phenomena. It assesses users' perception and knowledge by holding the expert diagram as the baseline. The Cox model asserts the need for an independent users' mental model through depicting a user-influence diagram as well as comparing the models of users or community members and those of experts. The present chapter explores methods of communicating climate change risk that are geared to the general public and/or users and emphasize information relevant to them. In order to introduce the issues of climate change–induced extreme environmental events and their associated risk, we first present an overview of the present discourse. We adopt a comparative approach to analyse the public and experts' beliefs about risks of floods, droughts, and heat waves in prairie communities. In addition, we offer a pioneering map of belief structures about risks of floods and droughts in rural communities and of heat waves in urban areas. Finally, we juxtapose public and expert understandings of risks to identify gaps in knowledge and miscommunications.

CLIMATE CHANGE, EXTREME WEATHER, AND DISASTER RISK

Climate Change and Environmental Extremes

We analyse extreme environmental events in a relative perspective of geophysical processes and events and look for remarkable deviation from the norm and potential harm to human and other lives, property, assets, and other resources (Haque and Burton 2005, 4; Smith

2006). Some observers write about increasing extremes in some parts of the world. These changes could be an outcome simply of natural variability but are also consistent with many of the shifts expected as greenhouse gases accumulate and the climate changes (Francis and Hengeveld 1998; O'Brien et al. 2008). Physical mechanisms could increase both the frequency and the intensity of extreme weather as a result of climate change. Global warming is highly likely to lead to a widespread increase in the amount of water and energy that moves through the hydrological system by increasing evaporation, transpiration, and the air's capacity to hold moisture (Francis and Hengeveld 1998). This scenario, combined with a more unstable atmosphere due to increased convection over warmer surfaces, is likely to increase the frequency and intensity of extreme weather (Francis and Hengeveld 1998; IPCC 2007).

The discourse on recent climate change and the correlation of increased variability of environmental extremes with global climate change have attracted stakeholders and policy- and decision-makers. Consensus on the relations between climate change and extreme events has not yet been reached among scholars because of the complex and non-testable nature of the problem. Recent research has suggested three interesting inferences.

First, the atmospheric thermal regime in the last few decades has risen consistently, particularly in the northern hemisphere (IPCC 2001; 2007). The IPCC report of 2007 asserts that warming during the past century is unequivocal because of observations of the rise in global average air and ocean temperatures, extensive melting of snow and ice, and rising global average sea level. Also, since about 1970, intense tropical cyclone activity has increased in the North Atlantic (IPCC 2007; Emanuel et al. 2008; Karim and Mimura 2008; Frank et al. 2010), with limited proof of increases elsewhere.

Second, there is much evidence that current mitigation and development policies and practices vis-à-vis climate change will allow global greenhouse gas emissions to continue expanding over the next few decades. Such an increase will alter temporal variability on all scales (daily, seasonal, inter-annual, and decadal) as well as the frequency, intensity, and duration of extreme events (IPCC 2001). Component features of the geophysical (e.g., avalanches, landslides) and climatic extremes (e.g., droughts, dry spells, hail storms, hurricanes, storms, tornadoes), however, exhibited mixed and thereby inconsistent trends, with considerable variations at different geographical scales.

Third, the IPCC Summary for Policymakers (IPCC 2007) asserts that confidence has grown among analysts relative to the earlier Third Assessment Report (TAR) in depicting projected patterns of warming and other regional-scale features, which include changes in wind patterns, precipitation, and some areas of extremes and sea ice. Regional-scale changes include "*very likely* increase in frequency of hot extremes, heat waves and heavy precipitation"; "*likely* increase in tropical cyclone intensity; less confidence in global decrease in tropical cyclone numbers"; "poleward shift of extra-tropical storm tracks with consequent changes in wind, precipitation and temperature patterns"; and "*very likely* precipitation increases in high latitudes and *likely* decreases in most subtropical land regions, continuing observed recent trends" (IPCC 2007, 8).

A clear correlation between global warming and environmental extremes in all areas cannot be established by current knowledge, although many features and regions exhibit extreme variability. Van Aalst (2006) argues that while the human emission of greenhouse gases is likely to continue to increase and the expected range of global temperature rise is between 1.4 and 5.8 Celsius degrees until 2100, there will be significant variations between regions.

Public versus Experts' Perception of Risk

Some observers have suggested a significant gap between public perception of risk and the risk communication strategies devised by expert groups through policy and planning. Risk perception can be defined as people's beliefs, attitudes, judgments, and feelings, as well as the broader social or cultural values and dispositions that people adopt, towards hazards and their benefits. By this definition, risk perceptions appear to be multidimensional and much more context-sensitive than formal measures of risk, which often address a single dimension (e.g., expected loss) (Pidgeon et al. 1992; Pidgeon 2008). There is one key difference: scientists usually define risk in terms of the effects on populations, while the public is concerned with the effects on individuals (Morgan et al. 2002).

The psychometric approach developed by Slovic and his co-workers (Slovic 1992; 2000) has indicated that different factors, such as whether the risk is perceived as involuntary, whether it will affect a large number of people, or whether it is seen to be unnatural, are likely to help determine the responses by residents and partly explain the disparity between their and experts' beliefs about risks. Also, the information system and the characteristics of public response that compose social amplification influence the nature and magnitude of risk (Kasperson et al. 1988; Boholm 2009). For some time now, discrepancies between expert assessments and residents or first responders' perceptions of risk – the "objective–perceived risk dichotomy" – have concerned risk managers and theorists (Slovic and Covello 1990).

Research has shown that people's risk perceptions and attitudes relate closely to the current level of risk reduction and to regulation employed to reduce risk (Lidskog 2008). However, experts often ignore the knowledge of a community audience. Their highly interconnected, extensive specific knowledge (Schimidt and Boshuizen 1992) makes it difficult for specialists to anticipate the public's wholly different perspective (Hinds 1999). Some observers explain such differences in terms of "rival rationalities," suggesting that residents look at risk more broadly than experts, whose knowledge is narrow and therefore likely to "miss something" of importance (Margolis 1997) – the "trap of the expert." Margolis attributes such stubborn conflicts less to what experts see that other people miss than to what ordinary people feel about risk that experts neglect (Margolis 1997). Experts tend to neglect the gap between their own knowledge and residents' (Hinds 1999), and they communicate in one direction – the "factual information" or "empty bucket" model.

Experts' perceptions of risk do not relate closely to any of the above-stated risk characteristics. Instead, experts appear to see riskiness as synonymous with expected annual mortality (Slovic, Fischhoff, and Lichtenstein 1979). Many conflicts about the acceptability of particular risks result from differing definitions of risk and resulting different assessments of the riskiness of an action or technology rather than from differences in opinion about acceptable levels of risk. The practical value of sensitivity to residents' perceptions is seen in the aura of enlightenment surrounding many excellent recent guides to risk communication (CRPC 1989).

A range of fundamental risk perceptions of environmental extremes and of issues about risk communication affects public response to discussions about the dangers of climate change (Dessai et al. 2004). Lorenzoni, Pidgeon, and O'Connor (2005), for example, report that Americans tend to view climate change as a moderate risk, affecting other populations or places, remote in space and time.

They also tend not to link climate change with direct health effects, indicating a clear gap between public and experts' risk assessments. Current literature on public opinion and knowledge concerning climate change suggests that the public lacks a clear understanding of the precise nature, causes, and consequences (Bostrom et al. 1994; Lorenzoni, Pidgeon, and O'Connor 2005; Weber 2006; Etkin and Ho 2007). In addition, the public commonly displays a variety of misunderstandings and confusion about the causes, and even very well-educated individuals tend to conceptualize climate change very differently from scientists and specialists. While systematic, scientific knowledge of the causes and dynamics of climate change is fundamental and necessary, it cannot capture the scope and interdisciplinary nature of the problem.

The IPCC's *Fourth Assessment Report 2007* underscores this concern, suggesting that, while the sciences should be the source of information and evidence on anthropogenic climate change and impact, the possible resulting danger will ultimately be judged by socio-political processes as well as in terms of aspects of uncertainty and risk. As a consequence of public lack of interest, misconception, and confusion and of specialists' failure to comprehend the broad interdisciplinary spectrum of climate change issues, wide gaps within and between the two groups do exist. Because there has been little communication concerning risk and uncertainty, gaps in risk communication have shaped the variation of knowledge and perception between them. Frewer (2004) asserts that traditional risk communication has been top-down and one-way: service providers convey experts' knowledge about risks, including environmental and health-related issues (Plough and Krimsky 1987). Such an approach tends to neglect public concerns or beliefs, leading to general distrust and lack of confidence in institutions, thus creating a gap between the public and experts' domains.

CLIMATE CHANGE AND DISASTER RISK IN MANITOBA: FLOODS, DROUGHTS, AND HEAT WAVE HAZARDS

Spanning more than 40 degrees of latitude and almost 100 of longitude, Canada occupies a vast geographical area. This characteristic, along with the effects of local and regional topography, means that climatologically extreme weather conditions are different for various urban centres (Bellisario 2001). Most disasters have been hydro-meteorological in origin (OCIPEP 2004), and the trend in disaster occurrence is rapidly shifting with climate change. Results from 37 weather stations, along with 50 sets of natural streamflow data and 13 sets of evapotranspiration data, have shown that the prairies have become warmer and drier over the last four to five decades (Gan 1998). Conforming to this view, Blair (1997) has observed that the average maximum and minimum temperatures in Winnipeg, Manitoba, have been rising (study period 1872–1993), supporting the assertion that a warmer global climate would result in less 1–15–day temperature variability. Canada's average annual temperature has also warmed by around 1C° over the last century – higher than the average global rise of 0.5C° (Zhang et al. 2000). In the southern regions of Canada, the average temperature rising by from 0.5C° to 1.5C° has been correlated with an increase in precipitation of 5 per cent and 35 per cent respectively during the last century (Zhang et al. 2000). The rapidity of the change also supports the hypothesis that climate changes tend to be non-linear.

By assessing the sensitivities and vulnerabilities of natural resources and human activities, in contrast to Blair's (1997) observation, Sauchyn and Kulshreshtha (2008) conclude that the most significant threat posed by climate change in the Canadian prairies is the projected increase in climate variability and frequency of extreme events. More specifically, rural and urban residents would face unique challenges concerning an increase in floods, drought, heat waves, and their impact on the socioeconomic, psycho-social, and physical well-being of communities and individuals. In this context, we outline the trends and patterns in floods, droughts, and heat waves in the region.

Several river basins in Manitoba are prone to flooding, of which the Red River valley is the principal one (Figure 9.1). The Red River basin in southern Manitoba occupies an extremely wide and flat floodplain that has been historically prone to severe floods. The 1997 flood was the largest recorded event of its kind in the 20th century in the Red River basin; the estimated unregulated discharge downstream of the Assiniboine River was 4,536 cu m per second (these two rivers meet in Winnipeg). The previous record-holder – the flood of 1950 – showed a considerably lower discharge (3,024 cu m per second) (Table 9.1). The largest recorded discharge through the Red River occurred in 1826: 6,300 cu m per second.

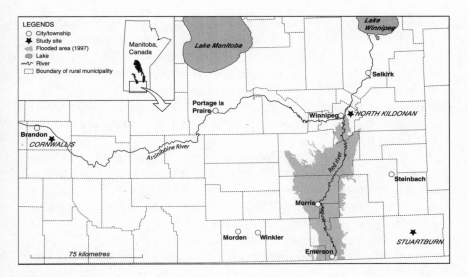

Figure 9.1
Location map of study sites (both urban and rural) and area flooded in 1997
by the Red River, Manitoba, Canada

Table 9.1
Flood flows (lowest–highest) and levels, James Avenue (assuming no flood
protection works in place), downtown Winnipeg, selected years

	Peak flow	James Avenue	
Year	Cubic metres/second	Metres above datum	Metres above sea level
1948	1,932	7.04	228.96
1966	2,480	8.02	229.77
1979	2,996	9.17	230.93
1950	3,024	9.24	230.99
1996	3,024	9.24	230.99
1861	3,500	9.81	231.57
1997	4,536	10.45	232.21
1852	4,620	10.52	232.27
1826	6,300	11.13	232.88

Source: Manitoba Water Commission. 1998. *An Independent Review of Actions Taken during the 1997 Red River Flood*, 14.

The 1997 flood required the evacuation of 28,000 people and caused an estimated $500 million in damage (Farlinger et al. 1998; Lemarquand 2007). The risk of flooding in the Canadian prairies is expected to worsen with climate change and the intensification of the hydrological cycle (IPCC 2007). Simonovic and Li (2004) found that climate change may increase annual discharge and shift ahead flood starting and peak occurrence in both the Assiniboine and the Red River basins.

Along with the threat of "too much water" at one time, "too little water" in a season has also historically threatened prairie communities. Many times in the twentieth century (i.e., 1936–8, 1961, 1976–7, 1980, 1984–5, and 1988) (Gan 2000), the region experienced periodic agricultural droughts, severely affecting both regional and national economies (Sauchyn 2004). Notably, in the 1930s, droughts affected 7.3 million ha of agricultural land and forced a quarter-million people to abandon the region (Godwin 1986). The 1984–5 drought affected agricultural production throughout the southern prairies and resulted in an estimated loss of over $1 billion in gross domestic product, or GDP (Ripley 1988). Similarly, the 1988 drought caused a loss of $4 billion in exports and forced 10 per cent of farmers from agriculture (Arthur 1988).

A rise in mean atmospheric temperature throughout the twentieth century has been correlated with an increase in drought (IPCC 2007). Accordingly, the temperature of the continental interior of North America is predicted to rise sharply and therefore increase the frequency, intensity, and scale of future Canadian prairie droughts (IPCC 2007; Wheaton et al. 2007). However, with more frequent and intense droughts, prairie agriculture may face risk that could overpower traditional buffers. The threat is significant, because cereal and grain production accounted for 82 per cent of cultivated land and 8.3 per cent of GDP in Canada in 2000 (Nyirfa and Harron 2002).

Heat wave hazards characteristically are quite different from floods and droughts and more catastrophic in terms of human morbidity and mortality (McMichael et al. 2003; Luber and McGeehin 2008), especially in urban settings. Heat wave disasters in recent years – for instance, in Athens in 1987, in India in 1988, in Chicago in 1995, and in northwest Europe in 2003 – killed thousands of people (e.g., 35,000 in Europe), especially the elderly and people who were already sick. These events evidently show the degree of vulnerability of cities.

Chapman (1995) notices that, despite the many deaths in Canada suspected to be heat-related, they are not labelled as such, mainly because of inconsistencies in cause-of-death reporting. Although extreme heat has increased deaths in some regions in Canada (Kalkstein and Smoyer 1993; Smoyer, Rainham, and Hewko 2000), this hazard has remained relatively unfamiliar and poorly recognized in the Canadian literature. The few publications reveal that – in terms of experiencing the highest temperatures and most frequent heat waves – the prairies, southern Ontario, and regions along the St Lawrence River valley in both Ontario and Quebec are particularly susceptible to heat waves (Smoyer-Tomic, Kuhn, and Hudson 2003).

A few studies have analysed heat wave hazard in terms of maximum temperature at or above 30–32°C in recent decades. Although Environment Canada (1996) defines a heat wave as "a period of more than three consecutive days of maximum temperature at or above 32° Celsius," Bellisario et al. (2001) apply a lower parameter (i.e., more than 30°C) for Canada as a whole. Table 10.2 illustrates the number of days in Winnipeg with a maximum temperature of more than 30°C during the period 1875–2005. As depicted in the trend line, there has been a steady increase in the number of days of extreme heat, particularly June–August (Figure 9.2). During the last 50 years, three times Winnipeg has experienced prolonged hot weather with maximum temperature more than 30°C for 25 days or more cosecutively. In contrast, during 1875–1960, such events occurred only once. The persistence of hot weather (a daily maximum temperature more than 30°C) for 15 to 20 days was also observed in Winnipeg – at least nine times during 1960–2005.

As elaborated by Smoyer-Tomic, Kuhn, and Hudson (2003), we distinguish between an analysis of temperature and heat wave characteristics and heat wave impacts. Because less extreme temperatures and shorter durations have harmed people (Kalkstein and Davis 1989; Smoyer, Rainham, and Hewko 2000), a different measurement has been suggested (i.e., humidex, by Masterton and Richardson 1979). Heat wave effects are influenced by several characteristics, which include their frequency (both in a given summer and a number of events during a longer period), duration, and intensity and daily minimum temperatures, as well as age characteristics of a population structure.

As Lemmen and Warren (2004) note, projected changes in climate may pose a range of challenges to Canada, especially in health,

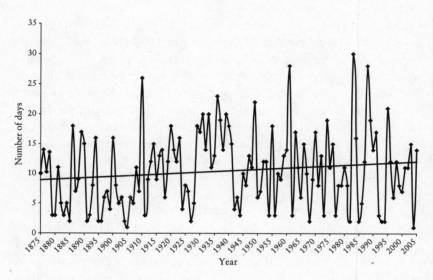

Figure 9.2

Number of hot days (daily maximum temperature >= 30°C) in Winnipeg, 1875–2005

Source: Modified from L. Bellisario et al. 2001 and L.M. Bellisario 2001.

Table 9.2

Heat wave occurrence in Canada, 1943–1998

City	Province	Heat wave counts				No. of events/ years of record
		2 days	3 days	4 days	5 days	
Vancouver	BC	2	0	0	0	0.04
Edmonton	AB	32	11	2	1	0.84
Calgary	AB	40	10	4	2	1.02
Saskatoon	SK	84	40	20	14	2.87
Winnipeg	MB	91	49	25	19	3.35
Ottawa	ON	80	42	26	28	3.20
Toronto	ON	79	45	16	26	3.02
Montreal	QC	63	24	14	22	2.51
Halifax	NS	4	1	0	0	0.09

Source: Data from L. Bellisario et al. 2001.

infrastructure, water, and other natural resources. A close relationship between climate and human health is viewed in the impact of extreme climate events and weather disasters. Flooding, droughts, heat waves, severe storms, and other climate-related environmental disasters can damage health and social well-being by increasing the risk of injury, illness, stress-related disorders, and death. There is a serious concern that climate change of the magnitude projected for the twenty-first century by the IPCC may have profound consequences for health and health care in Canada.

METHOD

To overcome the shortcomings of traditional risk communications and close the gap between experts and the public, cognitive scientists have used mental models to understand what humans know and perceive and how they decide and behave in uncertain situations. Although Kenneth Craik introduced the term in 1943 while studying cognitive mechanisms of learning processes in a system, interest in the application of mental models to risk and uncertainty grew during the 1980s and 1990s (Gentner and Steven 1983; Johnson-Laird 1983; Bostrom and Lash 2007). Craik (1943) postulated that the mind constructs small-scale models of reality and applies them to anticipate events, to reason, and to substantiate explanation.

Based on Craik's ideas, many cognitive scientists have insisted that the mind does create knowledge models as a means to construct cognition of our imagination, information, and comprehension. Johnson-Laird (1983) proposed mental models to describe the process whereby humans reason deductively, while Gentner and Steven (1983) proposed that they provide humans with information on how physical systems work. Analysts have begun using them in communicating about climate change because they predispose the mind towards specific ways of thinking about a problem, its causes, its effects, and its solution.

Analysts realized that warnings about climate change had triggered misunderstandings and inappropriate responses. Several studies revealed that people's perceptions and beliefs correspond to a variety of knowledge models about the issue, some of which mislead them regarding causes and solutions (Bostrom et al. 1994; Bostrom and Fischoff 2001; Sterman and Sweeney 2007; Moxnes and Saysel

2008). For example, many analysts attribute climate change to some extent to human activity, while others regard it as solely a natural atmospheric phenomenon. In general, knowledge models are useful in climate change research because they make clear the logical implications of explicit assumptions.

The research reported here uses a modified version of mental models that questions the conventional, one-way scheme (from expert to general public) and proposes an alternative. Craik's disciples have applied the models using a psychometric approach, focusing on the cognitive mechanisms of learning. In this chapter, we expand this framework to incorporate other pertinent dimensions of behaviour, cognition, and perception, particularly the role of cultural, economic, and social factors and social learning.

The term "mental model" does not capture the dimensions of social learning and other contextual conditions. We propose, building on Craik's mental models, a knowledge model that would represent both cognitive and social learning and enable making inferences about a system or problem and the mechanisms that affect both perception and behaviour (Borgman 1984). Also, the literature on critical social theories challenges the "hegemonic" (dominant) role of experts' knowledge (Geuss 1981; Lindlof and Taylor 2002) and calls for innovative approaches that recognize experts, the public, and other stakeholders simultaneously and respectfully.

In this study, we adopt a pluralistic knowledge approach, examining three selected climate change–induced environmental extremes, in order to analyse experts' and public knowledge and beliefs about such hazards. We mapped existing knowledge and beliefs about floods (rural), droughts (rural), and heat waves (urban). We chose these three phenomena to provide various extremes and settings – rural and urban. We expected that they would help us improve the general method and understand the diverse contexts. We anticipated that the different physical and social characteristics of the three types might affect how people communicate, understand, and act on information about risk, safety, and mitigation. For instance, they could vary in the possible mitigation measures, in their wider "safety culture," or in types and locations of residents. We therefore developed the public's knowledge models to capture some of the physical, social, and other complexities within these differing hazards.

The first step in applying the knowledge model involved the construction of the "general public influence diagram," through the iterative development of schematic diagrams that captured the current state of knowledge and the pooled beliefs of various groups concerning climate change–induced extreme environmental events. We applied the model separately to rural and urban areas. The rural municipalities (RMs) that participated in this research have had recent environmental disasters (i.e., floods and droughts) and were willing to take part in the research: Stuartburn and Cornwallis represent flood- and drought-vulnerable RMs, respectively, in Manitoba (Figure 9.1). Stuartburn is in southeastern Manitoba, 120 km south of Winnipeg, with 1,630 residents in 2005. Families had an average of $37,622 in annual income (national average: $44,136). Cornwallis, in southwestern Manitoba, lies in the province's agricultural heartland and grew during 2001–6, boosted by the relocation of the Kapyong Barracks from Winnipeg to Canadian Armed Forces Base Shilo in the southeast of the RM. The region boasts a farming economy; historically, residents have diversified their income via health care, social services, and business.

We interviewed 20 respondents – 10 from Stuartburn and 10 from Cornwallis – in March 2007, using open-ended survey instruments. The primary sampling units (PSUs) were households, and we interviewed the household heads to capture community and individual knowledge and beliefs about climate change–related extreme environmental events in their own words. We selected PSUs associated with agriculture or some other occupation in the community. We applied a "snowball" method to make contacts with participants able to read and write and willing to participate.

The focus of the urban sample was heat waves, and so, in selecting the community, we considered demographic composition and structure, housing conditions, socioeconomic status, state of social and health services, and – most important – vulnerability to heat waves. We chose the North Kildonan ward of Winnipeg: according to the 2001 census, 17.2 per cent of the 37,000 residents were 60 years or over and 80.1 per cent were 15 years or older, and the average household income was about $66,000, higher than the national average of $53,634 (Statistics Canada 2006). Historically, the north end of Winnipeg has contained low-income households and provided homes for working class and "foreign immigrants" (Carter 1997). At the time of the survey, the ward had mixed residential neighbourhoods, which included low-income housing complexes.

As the literature suggests, elderly and low-income groups are the most vulnerable during a heat wave (Applegate et al. 1981; Jones et al. 1982; Kenny et al. 2010); the sampling design therefore required the application of a stratified sampling procedure. We interviewed 25 respondents during February and March 2007. We obtained the sample from a list of households – a public-domain database – from the ward's councillor. A preliminary enquiry with his office revealed that he sought consent from residents to create such a public resource. We used systematic random sampling to select interviewees.

To fulfil the second step of the knowledge model process, we created quantitative confirmatory questionnaires and mailed them to a large sample of 400 PSUS in rural areas (200 in Stuartburn and 200 in Cornwallis). The response rate was 20.5 per cent in Stuartburn and 23 per cent in Cornwallis. In North Kildonan we mailed the questionnaire to 300 randomly selected PSUS. The response rate there was 38 per cent. We analysed some parts of the results with a five-point Likert scale, particularly to minimize measurement error and other biases.

In the third step, we constructed the "expert influence diagram" by iterative development of schematic diagrams that capture the current knowledge and pooled beliefs of experts concerning extreme events. We did this through an iterative, open focus-group discussion. The process enabled the creation of three separate models, with inputs from 12 experts from various organizations, including nongovernmental organizations, the federal and provincial governments, and research and academic institutions, who helped us to frame influence diagrams on the risks associated with climate change–induced floods, droughts, and heat waves. We selected leaders in their respective fields – climate change, disaster management, and health. These models illustrate the relationship of complex variables, with factors influencing each other in direct and indirect ways. We used them to identify certain factors and relationships that exist between these variables as well as to compare them with the public knowledge models.

By comparing expert and public knowledge, we attempted to develop a new, pluralistic approach, suggesting that risk communications would need a two- or multifaceted approach and that the expert model does not need to dictate the building of knowledge models. Rather, the juxtaposition of expert and public understandings is critical and can help us determine knowledge gaps and misunderstandings and reinforce appropriate beliefs and behaviour about safety and mitigation.

FINDINGS

Following both Morgan et al. (2002) and Cox et al. (2003), we developed the model of public knowledge using data from interviews and surveys and, to compare with it, the model of expert knowledge (influence diagrams). We adopted a new comparative approach to analyse the different expert and public beliefs about risks of floods, droughts, and heat waves in the prairie communities. In addition, our results mapped belief structures regarding risks associated with rural floods and droughts and urban heat waves, which previous studies did not fully attempt. Finally, the juxtaposition of expert and public understandings of risks enabled us to identify knowledge gaps and miscommunications; we analyse these in the discussion section.

Nature of the Public's Knowledge Model

To develop the general public's knowledge models, we borrowed a mixed method from both the mental model and the Delphi process. The original method for mental models at Carnegie Mellon applies the expert influence diagram as a structural template for analysing data from community members or users. Applying the idea-generating strategy (Linstone and Turoff 1975; Needham and de Loe 1990) of the Delphi process, we attempted to retain a strong contextual element, in which community members' understandings and concerns shape the public model. We structured the interview protocols to capture elicited views of various vulnerable groups and the general public with a minimum of prompting from the researchers or influence from other family or community members. The interviews took place in a private setting at home or community centres for 60–75 minutes, and we audio-recorded them with the agreement of the participants.

Table 9.3 contrasts the characteristics of public knowledge for the three environmental extremes. Four themes emerged, and we applied them to categorize public views, experiences, and beliefs: understanding of the physical processes; knowledge and perception of flood, drought, and heat wave risk; concerns about physical,

Table 9.3
Comparison of public model and expert influence diagrams, with gaps in knowledge and perception

Expert influence diagrams	Public knowledge models	Gaps in knowledge and perception
Climate change physical processes: GHG concentration and rise of temperature		
Human activities increasing GHGs, resulting in warming of atmosphere	Many people disbelieving or unaware of positive correlation between GHGs and warming of mean atmospheric temperature	Experts know physical processes, understand causes and effects; public has diverse and broader perspectives.
Rise in atmospheric mean temperature changing hydrological conditions and increasing floods	Misconceptions concerning physical processes: for example, many believing ozone depletion increasing mean atmospheric temperature	Public generally less interested in understanding physical processes and less certain about cause and effect
Increase in atmospheric mean temperature causing deficit in precipitation and soil moisture, leading to dry spells and droughts		
Rise in mean temperature causing heat waves		
Knowledge and perception of flood, drought, and heat wave risk		
Risk viewed in terms of probability of occurrence and vulnerability of people and resources; quantitative, precise methods applied	Risks to climate change–induced extreme environmental events seen in personal terms: problems in immediate surroundings and experiential learning	Risk indicators and measurements vary significantly between experts and the public.
	Hazard risk calculated in qualitative terms and relying heavily on previous experience	Experts' knowledge of risk more precise, focused, and converging on a specific topic; public knowledge more comprehensive, collective, and divergent

Table 9.3 (Continued)

Expert influence diagrams	Public knowledge models	Gaps in knowledge and perception
Concern about effects of flood, drought, and heat wave risk		
Flood: changes in biophysical environment and socioeconomic and demographic variables, physical damage to infrastructure, loss of property and capital, and psychological stress	Effects considered in personal terms rather than socioeconomic, infrastructural, and environmental aspects	Experts' concerns about effects of climate change–induced environmental extremes systematic and concentrating on various socioeconomic sectors as well as bio-physical environment
Drought: deterioration of bio-physical environment, constraints in water supply, loss of crops, and psychological stress	Urban residents concerned more about health effects of heat waves	Public concern about effects more personal
Heat wave: worsening of physiological and other health conditions, environmental damage (e.g., water and food contamination), infrastructural damage, and psychological stresses	Local flood effects not well understood by institutions	Immediate effects receiving more attention from public than experts' consideration of long-term impact
Knowledge of effects: systematic perspectives and sectoral in nature	Household-level flood effects of more concern than regional impact	
	Drought's impact on crops and family income main rural concern	
	Rural residents concerned more about psychological stress from drought's costs	
Knowledge of prevention and mitigation		
Climate change–induced extreme environmental events modifiable by intervention in physical process	Role of individuals in preventing and mitigating physical processes or impacts of climate change–induced environmental extremes nominal	Expert knowledge model: nominal explanations for individual behaviour change and emphasis on structural engineering interventions in physical systems
Risk reducible by change in behaviour	Institutional interventions and regulatory assertiveness more important than individual behaviour in dealing with risk from climate change–induced extreme environmental events	Public model: individual behaviour and actions in preventing and mitigating climate change–induced environmental extremes not viewed as significant
Institutional reforms necessary to improve risk communications and other interventions		

socioeconomic, and environmental effects; and knowledge of prevention and mitigation measures.

Understanding of Climate Change Physical Processes: GHG Concentration and Rise of Temperature

The general public and the various social groups consider that the physical processes concerning climate change and its effects on environmental extremes are correlated; such extreme events directly affect humans and tangible resources by creating chain effects. This view contrasted with expert experiences, which focused primarily on cause and effect. Although public knowledge reveals very diverse perspectives, many people (almost 40 per cent of the sample for floods and droughts) believed that global warming was not linked with greenhouse gas emissions resulting from human activities. Awareness of greenhouse gases per se appeared low: "Really, I don't think it's affected anything. In fact the Green House Gases, your trees would use up most of it" (Stuartburn no. 6 – senior male).

In the confirmatory questionnaire survey, why the earth's temperature is rising was explored among the general public. The elicited results reveal that, despite awareness of the relationship between human activity and the incremental rise in global temperature, there exist some misconceptions about determining factors. Many respondents thought that ozone depletion is causing global warming. In Morgan et al. (2002), US community respondents treated ozone depletion and global warming as synonymous.

We infer that many people do not perceive climate change and its associated features with clarity. Some rely heavily on experiential learning and exposure to various symptoms rather than understanding the relationships between variables.

Knowledge and Perception of Risk

The risk of various human spheres to environmental extremes is linked with the probability of occurrence of an event and the vulnerability of humans and their resources (Wisner et al. 2004; Smith 2006). Our public knowledge model reveals that knowledge and perception of climate change–induced environmental extremes vary by type of hazard and associated risk. Among the rural sample, the risk of crop loss and the risk of homesteads and other socioeconomic loss are well conceptualized, whereas most members (82.4 per cent) of the urban sample (North Kildonan) believed that Winnipeg was unlikely to experience a heat wave in the next 5–10 years or were unaware of such a risk. This was perhaps the result of both poor conceptualization of what a heat wave is and ignorance about its probability. Probabilistically, this perceptual configuration contrasts sharply with observed temperatures and heat waves (see Bellisario et al. 2001).

Although experiential knowledge shapes public perception of the risk of climate change–induced environmental extremes, a complex set of factors affects the configuration of public knowledge and perception. While experts tend to use past occurrences to calculate probabilities of future events and potential losses, the public perceives risk in terms of individual or collective interest. Several explanations have emerged about why various people perceive the risk of types of hazards in a varied manner (Weber 2001). The psychometric paradigm helps explain people's extreme aversion to some hazards, their indifference to others, and the discrepancies between these reactions and experts' opinions. People make quantitative judgments about the current and desired riskiness of diverse hazards and the desired level of regulation of each. These judgments are then linked to other properties of decision-making.

Our interviews show that community members in general cannot escape from the frame of experiential knowledge-gathering. However, the configuration of risk maps is affected by a large set of variables relating to individual values, belief systems, and personal and socio-cultural contexts.

Concern about the Effects of Hazards

Public concerns differ for floods, droughts, and heat waves. For floods, expressed concerns were relatively high, mainly in terms of loss of property and other resources and psychological stress. Human vulnerability to flooding was recognized by about 84 per cent of Stuartburnites. Similarly, Cornwallisites displayed concern over the effects of increased temperature and droughts on soil moisture deficit and their reduction of agricultural production.

Among the Winnipeggers, there appeared a general acknowledgment of infrastructural and environmental effects, and people worried about the potential health effects of heat waves, but most did not think themselves at immediate risk. They worried more about individual loss potential rather than societal loss. The elderly and poor expressed serious concern about health effects, but the former worried more about their gardens, although non-gardeners did not express such concern.

Knowledge of Prevention and Mitigation of Hazards

Research has shown that people view risk and hazard-related issues from a much broader perspective than experts. Usually, their risk perceptions and attitudes relate closely to current measures of reducing risk, such as – for flooding – the construction of diversion channels, embankments, ring dikes, and other structural engineering schemes. Similarly, for heat waves, respondents cite heat alarm systems and the rationing of power and advocate the strict application of existing regulations to reduce risk.

What an individual and society at large can do about preventing, reducing, and coping with climate change and its associated environmental extremes is pivotal to addressing issues and problems. Interviewees raised specific issues, particularly to delineate their ideas on prevention and mitigation at various scales: individual, community, and institutional. They showed that many people have good and considerable knowledge about reducing emissions. Echoing a collective response, one commented: "Adaptive behaviour is necessary and I think this involves recycle, reuse and conservation … Public transport needs to be improved by government so that people don't need to use car to go to remote places or even government motivation is necessary to enforce emission control laws."

The results of our interviews showed that they would prefer stricter regulation, particularly enforcement of laws and regulations to reduce emissions and thereby address global warming. These findings suggest heavy reliance on institutional initiative and implementation vis-à-vis large-scale problems, particularly in areas with considerable uncertainty.

Experts' Knowledge Models

We capture the experts' knowledge models by influence diagrams on flooding, droughts, and heat waves. These directed networks (Atman et al. 1994) use arrows or "influences" to connect related "nodes" (Morgan et al. 2002). Influence diagrams work for knowledge models because they are applicable to virtually any risk, are compatible with conventional scientific/technical ways of thinking, are easy to understand and subject to peer review, and fit within a decision-making perspective (Morgan et al. 2002).

Experts developed an influence diagram for each type of event based on their beliefs, attitudes, and values. Each diagram emerged in four stages: a review of the literature on types of events; content analysis of existing information on the physical process, state of risk, prevention, and mitigation; focus group discussions with experts; and validation of the diagram by a group of external experts.

Because this process is interdisciplinary, with inputs from experts on various aspects of the thematic areas, the outputs by nature remain provisional (Cox et al. 2003). In our focus group discussions, we presented our initial influence diagram (from the first two stages) for review by these experts. We added their comments in the iterative development of the diagrams. We also invited them to develop comprehensive influence diagrams on each hazard and then asked a group of external experts to review and validate the outputs, which resulted in the final diagrams in Figures 9.2 and 9.3. These diagrams are provisional or tentative, as many dynamic factors could affect their validity. The influence diagrams are inclusive – a comprehensive, collective representation of expert understanding of the risk of these extreme events.

Figure 9.3 (a and b) depicts experts' view of the sequential relationship that leads to risk associated with flooding. It is a generalized version of the expert knowledge model of floods, revealing factors and influences that lead from climate change to incremental flooding and eventually psychological stress. Each of the four parts of Figure 9.3a represents a domain of conditions: human activities and concentration of greenhouse gases (GHGs) (part 1), atmospheric and other physical conditions (part 2), risk and human vulnerability (part 3), and human/social effects (part 4). Part 3 (Figure 9.3b) shows the interface between physical exposure and human vulnerability to floods, which creates the risk to floods, with multiple facets of potential effects.

The experts' drought model (Figure 9.4a and b) follows a series of key relationships that show the factors that lead to the risk associated with agricultural drought. The model has three parts, each representing a different domain: atmospheric conditions (part 1), physical environment and the risk (part 2), and human/social effects

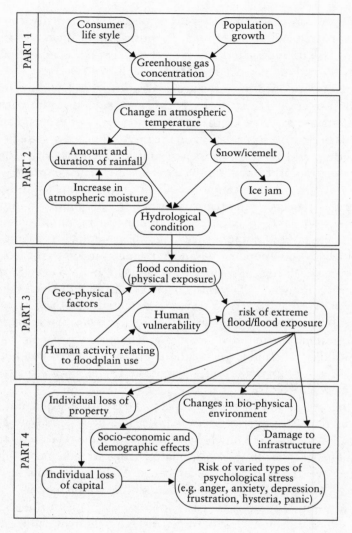

Figure 9.3a
Experts' knowledge model concerning climate change–induced floods

(part 3). Figure 9.4a is a generalized version of the expert knowledge model, which displays the major factors and influences that lead from climate change to drought and eventually affect individuals and households; we show details of only part 3 (Figure 9.4b).

The main factors influencing urban heat waves are a rise in atmospheric mean temperature, changes in surface atmospheric moisture,

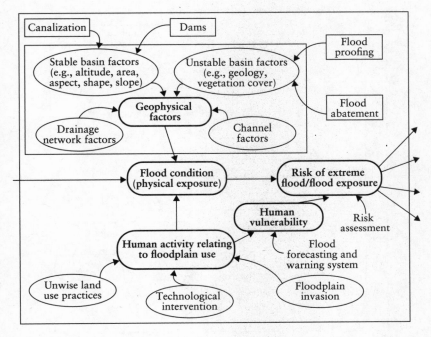

Figure 9.3b
Experts' knowledge model concerning climate change–induced floods (details to part 3 of Figure 9.3a)

and changes in atmospheric circulation. We determined that the rise in mean atmospheric temperature is considerably influenced by the GHG concentration in the atmosphere. Two variables – consumers' way of life and population growth – cause the increase in concentration. The model shows that the risk of extreme heat on human health is a function not only of heat waves but also of people's long exposure to sunlight, their lack of access to water, and poor housing, with inadequate ventilation.

DISCUSSION

Our study helps explain how experts and the public understand and communicate about climate change–induced extreme environmental events and generates first-order representations of experts' and public beliefs. We can see the contrast between the experts' relatively well-structured influence diagrams and public beliefs and

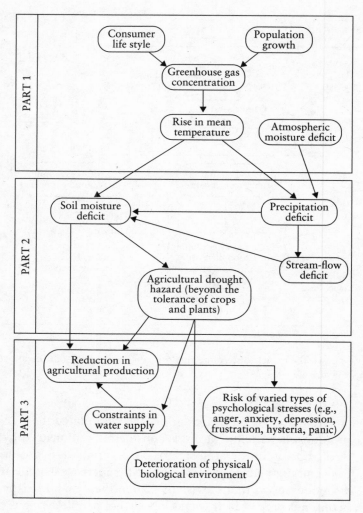

Figure 9.4a
Experts' knowledge model concerning climate change–induced droughts

understandings of the flood, drought, and heat wave hazards. As we showed above, our approach differs from the mental models of Morgan et al. at Carnegie Mellon University, where a template based on the initial expert influence diagram guides public knowledge and beliefs on those topics. We employ an "overlay" of the expert and public models. Our approach forgoes some of the direct first-order comparability to remain sensitive to the contextual details of local

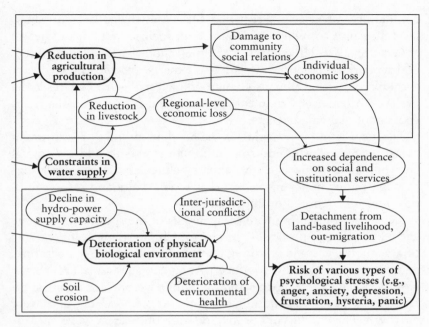

Figure 9.4b
Experts' knowledge model concerning climate change–induced droughts (details to part 3 of Figure 9.4a)

public knowledge and understanding, as well as to underlying contextual issues.

When expert and public knowledge models are developed with open-ended structures, it is still possible to compare them. Inferences in Table 9.3 synthesize the results of these comparisons. For the first two columns, we examined the essence of the findings in the two models. We began with the public model to structure this comparison. In dealing with each of the four themes that emerged from the public and its sub-categories (low-income and elderly), we considered the expert models for information relevant to both public models and the themes themselves. We noticed that the generic themes suggested by the four headings permit a systematic interrogation and comparison of the two types of models without suppressing the diversity of views. Specifically, we organized the comparison by the normative objective of determining the gaps, misunderstandings, and correct aspects of public beliefs on risk and mitigation, as well as contextual factors that limit appropriate action. For instance, our community interviewees reported little knowledge of individual-level measures to reduce environmental, socioeconomic, and health effects of hazards. In contrast, the expert models have identified several types of such actions. Relative to experts' specific knowledge of the links between climate change and global warming, residents draw on broader environmental and societal factors to make inferences about climate change.

The expert models further revealed that although institutional measures play a profound role in human efforts to address critical environmental issues (such as climate change and ozone depletion) through public policy measures, individuals' roles and responsibilities cannot be ignored or underestimated.

It is worth presenting here some details of the specific knowledge gaps. For physical processes concerning climate change, for the role of greenhouse gases in warming the mean atmospheric temperature, and for the relationships with environmental extreme events, the gaps between expert models and the public were wider. For risk, effects, and prevention and mitigation measures, knowledge gaps were narrower. The findings of our investigation of the heat wave hazards determined knowledge gaps in six areas: understanding the natural processes and climate change dynamics; comprehending the causes of the rise of global atmospheric temperature; conceptualizing heat waves, particularly in terms of the required duration to be a threat to life and property; identifying the impacts of heat waves other than on the health sector; heat wave "risk estimation" for Winnipeg and its neighbourhoods; and the role of precautionary measures at the individual level in affecting the potential mortality rate due to heat wave events. The most conspicuous gaps were in two areas: when we asked the public about heat wave "risk identification" and "risk estimate" in Winnipeg, their responses revealed their lack of recognition of the heat wave risk, their lack of interest in such an issue, or their lack of recent experience; and they were aware of the nature of the risk to heat exposure and of the potential adjustment of individuals' behaviour when exposed to hot weather, but they did not believe that the heat wave mortality rate could be minimized by personal or institutional precautionary measures.

On the one hand, the process underscores areas of appropriate public knowledge, public misconceptions, and gaps in public knowledge in relation to the expert model. These provide the basis for designing subsequent risk communications, programs, and policies. To take just one instance of each, both experts and the public emphasized the significance of climate change for extreme environmental events and the health risk to them (Table 9.3). In the case of the urban elderly, a common belief was that heat wave vulnerability is outside the individual's domain, and therefore individuals are not in a position to affect them or reduce impact by undertaking proactive measures. The identification of this knowledge gap can assist the expert model.

On the other hand, the public model may highlight areas of expert knowledge that require special attention. For example, the experts have limited understanding of the local context, the state of the problem, and appropriate intervention. While community members acknowledged the overarching importance of institutional and regulatory measures, they also noted that macro-level policies generally ignore locally needed resources and appropriate interventions.

CONCLUSIONS

We hope that the findings from our study can inspire policies that would lead to an increased emphasis on risk communication as a preventive approach towards climate change–induced extreme environmental events by increasing community coping capacity. Unfortunately, despite the overwhelming empirical evidence, climate change has been consistently either neglected or deferred as a priority issue, in both political and social realms.

In the context of reducing risk to climate change–induced extreme environmental events, other than mitigation, the most effective option is to reduce vulnerability (Haque and Burton 2005). This approach has, however, not yet been accepted in public policy, where extreme environmental events are still portrayed as "deviations from order of the established structure" (Hewitt 1983, 29). Risk and disaster management policies have thus traditionally propagated a distributive approach, in which adaptation strategies tend to be reactive, unevenly distributed, and focused on coping rather than prevention (Field et al. 2007). Although the vulnerability and resilience paradigm has weakened the disaster-response paradigm within the social science literature and many of the professional emergency and disaster management concepts, it has yet to affect policy and, more important, public discourse (Haque and Etkin 2007). In democratic nations, such as Canada, public awareness should ultimately coerce public policy into action. Sjoberg and Drottz-Sjoberg (2008), in this context, have noted that risk perception among the public and politicians is similar, in contrast with the experts. Such similarity can inform public policies on risk communication and regulation.

In attaining the social and environmental goals above, Cox et al.'s (2003) observation on the prioritization of the communication content is worth noting here. They asserted that informed judgment could be used initially to prioritize the climate change–induced, risk-related messages, and this should be supported through an iterative assessment of the communication. Suggestions from experts in the climate change and associated risk domains and analysis of resource users would support the effectiveness of the communication in offering the public information that it needs to protect itself and which is relevant to the livelihoods and progress of local communities.

Our study has provided a basis for developing general public-centred risk communications in the emerging areas of climate change–induced environmental risk. It has exhibited how an expanded knowledge model encapsulates considerable complexity and contextual information in public representations and understandings. In addition, the study varied from others by using mental models to inform risk communications that have emphasized a single risk issue or hazard. By pursuing a comparative study, we have been able to show that the method reveals very different concerns for three different hazards. We hope that by recognizing the value of the knowledge and information generated from public-centred knowledge models, the concerned public and private institutions will modify their risk communication methods and messages concerning climate change–induced disaster risk.

ACKNOWLEDGMENTS

The authors gratefully acknowledge the financial support of the Earth Science Division, Natural Resources Canada, and the Social Science and Humanities Research Council (SSHRC), Ottawa. We are also grateful to David Etkin, York University, and Brenda Murphy, Wilfrid Laurier University, for their comments and criticisms on an earlier draft of this chapter.

REFERENCES

Applegate, W.B., J.W. Runyan, Jr, L. Brasfield, M.L. Williams, C. Konigsberg, and C. Fouche. 1981. "Analysis of the 1980 heat wave in Memphis." *Journal of the American Geriatric Society* 29: 337–42.

Arthur, L.M. 1988. "The implications of climate change for agriculture in the prairies." *Climate Change Digest* 88–01: 1–11.

Atman, C.J., A. Bostrom, B. Fischhoff, et al. 1994. "Designing risk communication: completing and correcting mental models of hazardous processes, part I." *Risk Analysis* 14 (5): 779–88.

Bellisario, L., et al. 2001. *Assessment of Urban Climate and Weather Extremes in Canada – Temperature Analysis. Final Report to Emergency Preparedness Canada.* Ottawa: Emergency Preparedness Canada.

Bellisario, L.M. 2001. "Temperature extremes in Canadian urban centres: when and where are we at risk?" *Canadian Meteorological and Oceanographic Society 35th Congress.* Winnipeg.

Blair, D. 1997. "Short period temperature variability at Winnipeg, Canada, 1872–1993: characteristics and trends." *Theoretical and Applied Climatology* 58 (3–4): 147–59.

Boholm, A. 2009. "Speaking of risk: matters of context." *Environmental Communication* 3 (3): 335–54.

Borgman, C.L. 1984. *The User's Mental Model of an Information Retrieval System: Effects on Performance.* Ann Arbor, MI: University Microfilms International, Ann Arbor.

Bostrom, A., C. Atman, B. Fischoff, and M.G. Morgan. 1994. "Evaluating risk communications: completing and correcting mental models of hazardous processes." *Risk Analysis* 14: 789–98.

Bostrom, A., and B. Fischoff. 2001. *Communicating Health Risks of Global Climate Change.* New York: Emerald Group Publishing Limited.

Bostrom, A., and D. Lashof. 2007. *Weather or Climate Change in Creating a Climate for Change: Communicating Climate Change.* New York: Cambridge University Press.

Carter, T. 1997. "Winnipeg: heartbeat of the province." In J.E. Welsted and C. Stadel, eds., *The Geography of Manitoba: Its Land and Its People.* Winnipeg: University of Manitoba Press: 136–51.

Chapman, B. 1995. "When is death heat-related?" *CAP Today:* 56–61.

Cox, P., et al. 2003. "The use of mental models in chemical risk protection: developing a generic workplace methodology." *Risk Analysis* 23 (2): 311–24.

Craik, K.J.W. 1943. *The Nature of Explanation*. Cambridge: Cambridge University Press.

CRPC (Committee on Risk Perception and Communication). 1989. *Improving Risk Communication*. Washington, DC: National Academy Press.

Dessai, S., W.N. Adger, M. Hulme, J. Turnpenny, J. Kohler, and R. Warren. 2004. "Defining and experiencing dangerous climate change: an editorial essay." *Climate Change* 64: 11–25.

Emanuel, K.A., et al. 2008. "Hurricanes and global warming: results from downscaling IPCC AR4 simulations." *Bulletin of the American Meteorological Society* March: 347–67.

Environment Canada. 1996. *Climate and Weather Glossary of Terms*. Downsview, ON: Climate and Water Systems Branch.

Etkin, D., and E. Ho. 2007. "Climate change: perceptions and discourses of risk." *Journal of Risk Research* 10 (5): 623–41.

Farlinger, D., L. Chambers, F. Beaudette, D. Burn, and M. Hodgson. 1998. *An Independent Review of Actions Taken during the 1997 Red River Flood*. Winnipeg: Manitoba Water Commission.

Field, C.B., L.D. Mortsch, M. Brklacich, D.L. Forbes, P. Kovacs, J.A. Patz, S.W. Running, and M.J. Scott. 2007. "North America. Climate change 2007: impacts, adaptation and vulnerability." In M.L. Parry, J.P. Palutikof, P.J. van der Linden, and C.E. Hanson, eds., *Contribution of Working Group II to the Fourth Assessment Report of the Intergovernmental Panel on Climate Change*. Cambridge: Cambridge University Press: 617–52.

Francis, D., and H. Hengeveld. 1998. *Extreme Weather and Climate*. Downsview, ON: Environment Canada. Published by authority of the minister of environment.

Frank, D.C., et al. 2010. "Ensemble reconstruction constraints on the global carbon cycle sensitivity to climate." *Nature* 463: 527–30.

Frewer, L. 2004. "The public and effective risk communication." *Toxicology Letters*: 391–7.

Gan, T.Y. 1998. "Hydroclimatic trends and possible climatic warning in the Canadian prairies." *Water Research* 34 (11): 3009–15

– 2000. "Reducing vulnerability of water resources of Canadian prairies to potential droughts and possible climatic warming." *Water Resources Management* 14: 111–35.

Gentner, D., and A.L. Steven. 1983. *Mental Models*. Hillsdale, NJ: Lawrence Erlbaum Associates Inc. Publishers.

Geuss, R. 1981. *The Idea of a Critical Theory*. Cambridge: Cambridge University Press.

Godwin, R.B. 1986. "Drought: a surface water perspective." In *Drought: The Impending Crisis? Proceedings of the Canadian Hydrology Symposium No. 16*: 27–43.

Haque, C.E., and I. Burton. 2005. "Adaptation options strategies for hazards and vulnerability mitigation: an international perspective." *Mitigation and Adaptation Strategies for Global Change* 10: 335–53.

– and D. Etkin. 2007. "People and community as constituent parts of hazards: the significance of societal dimensions in hazards analysis." *Natural Hazards* 41 (2): 271–82.

Hewitt, K. 1983. *Interpretations of Calamity: From the Viewpoint of Human Ecology*. Boston: Allen and Unwin Inc.

Hinds, P.J. 1999. "The curse of expertise: the effects of expertise and debiasing methods on predictions of novice performance." *Journal of Experimental Psychology: Applied* 5: 205–21.

IPCC (Intergovernmental Panel on Climate Change). 2001. *Climate Change 2001: Synthesis Report – Summary for Policy Makers*. Wembley, England.

– 2007. *The AR4 Synthesis Report (IPCC Fourth Assessment Report)*. Valencia, Spain: Intergovernmental Panel on Climate Change.

Johnson-Laird, P.N., ed. 1983. *Conditionals and Mental Models: On Conditionals*. Cambridge: Harvard University Press.

Jones, T.S., A.P. Liang, E.M. Kilbourne, M.R. Griffin, P.A. Patriarca, S.G. Wassilak, et al. 1982. "Morbidity and mortality associated with the July 1980 heat wave in St. Louis and Kansas City." *Journal of the American Medical Association* 247: 3327–31.

Kalkstein, L.S., and R.E. Davis. 1989. "Weather and human mortality: an evaluation of demographic and interregional responses in the United States." *Annals of the Association of American Geographers* 79: 44–64.

Kalkstein, L.S., and K.E. Smoyer. 1993. *The Impact of Climate on Canadian Mortality: Present Relationships and Future Scenarios*. Canadian Climate Centre Report No. 93-7. Downsview, ON: Ontario Atmospheric Environment Service, Canadian Climate Program.

Karim, M.K., and N. Mimura. 2008. "Impacts of Climate change and sea-level rise on cyclonic storm surge floods in Bangladesh." *Global Environmental Change* 18 (3): 490–500.

Kasperson, R.E., et al. 1988. "The social amplification of risk: a conceptual framework." *Risk Analysis* 8 (2): 177–87.

Kenny, G., J. Yardley, C. Brown, R. Sigal, and O. Jay. 2010. "Heat stress in older individuals and patients with common chronic diseases." *Canadian Medical Association Journal* 182 (10): 53–60.

Lemarquand, D. 2006. "Red River flooding: mitigation planning in an international river basin." In J. Schanze, E. Zeman, and J. Marsalek, eds., *Flood Risk Management: Hazards, Vulnerability and Mitigation Measures*. Dordrecht, Netherlands: Springer: 207–18.

Lemmen, D.S., and F.J. Warren, eds. 2004. *Climate Change Impacts and Adaptation: A Canadian Perspective*. Ottawa: Climate Change Impacts and Adaptation Directorate, Natural Resources Canada.

Lidskog, R. 2008. "Scientised citizens and democratised science: reassessing the expert–lay divide." *Journal of Risk Research* 11 (1): 69–86.

Lindlof, T.R., and B.R. Taylor. 2002. *Qualitative Communication Research Methods*. 2nd ed. Thousand Oaks, CA: Sage Publications.

Linstone, H.A., and M. Turoff. 1975. "Introduction." In H.A. Linstone and M. Turoff, eds., *The Delphi Method: Techniques and Applications*. Reading, MA: Addison-Wesley Publishing Co.: 3–12.

Lorenzoni, I., N.F. Pidgeon, and R.E. O'Connor. 2005. "Dangerous climate change: the role for risk research." *Risk Analysis* 25(6): 1387–97.

Luber, G., and M. McGeehin. 2008. "Climate change and extreme health events." *American Journal of Preventive Medicine* 35 (5): 429–35.

Margolis, H. 1997. *Dealing with Risk: Why the Public and the Experts Disagree on Environmental Issues*. Chicago: University of Chicago Press.

Masterton, J.M., and F.A. Richardson. 1979. *Humidex: A Method of Quantifying Human Discomfort due to Excessive Heat and Humidity*. Environment Canada, CLI 1–19. Downsview, ON: Atmospheric Environment Service.

McMichael, A.J., et al., eds. 2003. *Climate Change and Human Health: Risks and Responses*. Geneva: World Health Organization.

Morgan, M.G., B. Fischhoff, A. Bostrom, and C. Alman. 2002. *Risk Communication: A Mental Model Approach*. Cambridge University Press.

Moser, S.C. 2007. "More bad news: the risk of neglecting emotional responses to climate change information." In S. Moser and L. Dilling, eds., *Creating a Climate for Change: Communicating Climate Change and Facilitating Social Change*. Cambridge: Cambridge University Press: 64–80.

Moxnes, E., and K. Saysel. 2008. "Misperceptions of global climate change: information policies." *Climatic Change* 93 (1): 15–37.

Needham, R.D., and R.C. de Loe. 1990. "The policy Delphi: purpose, structure and application." *Canadian Geographer* 34 (2): 133–42.

Nyirfa, W.N., and B. Harron. 2002. *Assessment of Climate Change on the Agricultural Resources of the Canadian Prairies*. Regina: Agriculture and Agri-foods Canada, Prairie Farm Rehabilitation Administration.

O'Brien, K., L. Sygna, R. Leichenko, W.N. Adger, J. Barnett, T. Mitchell, L. Schipper, T. Tanner, C. Vogel, and C. Mortreux. 2008. *Disaster Risk Reduction, Climate Change Adaptation and Human Security*. Oslo: Global Environmental Change and Human Security.

OCIPEP (Office of Critical Infrastructure Protection and Emergency Preparedness). 2004. "Tornado information." Ottawa: Office of Critical Infrastructure Protection and Emergency Preparedness. DOI: www.ocipep.gc.ca/disaster/default.asp

Pidgeon, N. 2008. "Risk, uncertainty and social controversy: from risk perception and communication to public engagement." In G. Bammer and M. Smithson, eds., *Uncertainty and Risk: Multidisciplinary Perspectives*. London: Cromwell Press: 349–60.

Pidgeon, N., C. Hood, D. Jones, B. Turner, and E. Gibson, eds. 1992. "Risk perception." In *Risk: Analysis, Perception and Management*. London: Royal Society: 89–134.

Plough, A., and S. Krimsky. 1987. "The emergence of risk communication studies: social and political context." *Science, Technology, and Human Values*: 4–10.

Ripley, E.A. 1988. *Drought Prediction on the Canadian Prairies*. Saskatoon, SK: National Hydrology Research Centre.

Sauchyn, D. 2004. *A 250 Year Climate and Human History of Prairie Drought*. Regina, SK: Canadian Plains Research Centre, University of Regina.

Sauchyn, D., and S. Kulshreshtha. 2008. "Prairies." In D.S. Lemmen, F.J. Warren, J. Lacroix, and E. Bush, eds., *From Impacts to Adaptation: Canada in a Changing Climate 2007*. Ottawa: Government of Canada: 275–328.

Schmidt, H.G., and H.P.A. Boshuizen, eds. 1992. *Encapsulation of Biomedical Knowledge: Advanced Models of Cognition for Medical Training and Practice*. Berlin: Springer.

Simonovic, S.P., and L. Li. 2004. "Sensitivity of the Red River basin flood protection system to climate variability and change." *Water Resources Management* 18: 89–110.

Sjoberg, L., and B. Drottz-Sjoberg. 2008. "Risk perception by politicians and the public." *Energy and Environment* 19 (3): 455–83.

Slovic, P. 2000. *Perception of Risk*. London: Earthscan.

– 1992. *Perception of Risk: Reflections on the Psychometric Paradigm. Social Theories of Risk*. New York: Praeger.

Slovic, P., B. Fischhoff, and S. Lichtenstein. 1979. "Rating the risks." *Environment* 21 (3): 14–20, 36–9.

Slovic, P.K., and V.T. Covello. 1990. "What should we know about making risk comparison?" *Risk Analysis* 10 (3): 389–92.

Smith, K. 2006. *Environmental Hazards: Assessing Risk and Reducing Disaster*. London: Routledge.

Smoyer, K.E., D.G.C. Rainham, and J.N. Hewko. 2000. "Heat stress mortality in five cities in southern Ontario: 1980 to 1996." *International Journal of Biometeorology* 44: 190–7.

Smoyer-Tomic, K., R. Kuhn, and A. Hudson. 2003. "Heat wave hazards: an overview of heat wave impacts in Canada." *Natural Hazards* 28: 463–85.

Statistics Canada. 2006. Population Census of Canada, 2001. www.statcan.ca/start.html

Sterman, J., and L. Sweeney. 2007. "Understanding public complacency about climate change: adults' mental models of climate change violate conservation of matter." *Climatic Change* 80 (3): 213–38.

Van Aalst, M.K. 2006. "The impact of climate change on the risk of natural disasters." *Disasters*: 5–18.

Weber, E.U. 2001. "Personality and risk taking." In N.J. Smelser and P.B. Baltes, eds., *International Encyclopaedia of the Social and Behavioural Sciences*. Oxford: Elsevier Science Limited: 11274–6.

– 2006. "Experience-based and description-based perceptions of long-term risk: why global warming does not scare us yet." *Climate Change* 77 (1–2): 103–20.

Wheaton, E., G. Koshida, B. Bonsal, T. Johnston, W. Richards, and V. Wittrock. 2007. *Agricultural Adaptation to Drought (AAD) in Canada: The Case of 2001 to 2002*. Saskatoon: Saskatchewan Research Council.

Wisner, B., P. Blaikie, T. Cannon, and I. Davis. 2004. *At Risk: Natural Hazards, People's Vulnerability and Disasters*. New York: Routledge.

Zhang, X., L.A. Vincent, W.D. Hogg, and A. Niitsoo. 2000. "Temperature and precipitation trends in Canada during the 20th century." *Atmosphere-Ocean* 38 (3): 395–429.

10

Natural Hazard Identification, Mapping, and Vulnerability Assessment in Atlantic Canada: Progress and Challenges

NORM CATTO

INTRODUCTION

Atlantic Canada is commonly perceived by both its residents and others to have relatively few natural hazards. Even though the Burin Tsunami of 1929 and its death toll of 27 people in Newfoundland and one in Cape Breton (see Ruffman 1991; 1993; 1995b) is regularly recalled on its anniversary of 18 November, few Newfoundland residents outside the communities directly affected appear to remember the event until it is brought to their attention. Repeated informal surveys and conversations between 1989 and 2010 in St John's high schools, and in Memorial University classes ranging from introductory courses to graduate school, in disciplines varying from geography through environmental science and earth science to education, indicate that few students consider Atlantic Canada intensely affected by natural hazards: events such as Hurricanes Juan in 2003 and Igor in 2010 are seen as highly anomalous exceptions (Catto and Tomblin 2009a; 2009b; in press). The lack of volcanoes and towering mountains and the rareness of terrestrial-focus earthquakes appear to have translated to a belief that the region has few "significant" natural hazards. The fortunately limited death tolls of some of the largest natural disasters, such as the 28 people who died in the Burin Tsunami, also lessen their prominence with residents.

Historical natural disasters, such as the "Great Independence Hurricane" of 12–16 September 1775, which killed a large but undetermined number of people in eastern Newfoundland and St-Pierre-et-Miquelon (possibly as many as 4,000; see Stevens and Staveley 1991; Ruffman 1995a; 1996; Stevens 1995; Liverman, Catto, and Batterson 2006); the "Great Miramichi Fire" in central New Brunswick in 1825, which killed approximately 200 people and burned 25–30 per cent of the province (Murdoch 1825; Marsh 1864); the Burin Tsunami; or the destruction of La Manche, NL, in 1966 by storm surge (Catto, Scruton, and Ollerhead 2003), are much less well known than human-induced events such as the Halifax Explosion in 1917, the Springhill Mine (1958) and Westray Mine (1992) disasters, or the torpedoeing of the ferry *Caribou* in 1942. Disasters that resulted from weather events, such as the loss of the *Ocean Ranger* in 1982, are also commonly considered human-induced rather than natural occurrences. Even residents who have knowledge of a single natural hazard in their community are generally unaware of the range, variety, or regional prevalence of hazards facing Atlantic Canadians (Catto 2008; Catto and Tomblin 2009a; 2009b; in press).

The reality, however, is that the region is subject to a variety of natural hazards (Figures 10.1a and b). Hurricanes and associated extratropical transitions, late autumn and winter storms, storm surges, ice storms and consequent power blackouts (e.g., St John's in December 1994; southeastern New Brunswick in January 1998; Halifax in March 2001 and November 2004; Bonavista in 2008 and 2010), extremes of heat and cold, droughts (e.g., Annapolis Valley in summer 1997; central Newfoundland in summer 2006), blizzards, and excessive seasonal cumulative snowfalls (e.g., 650 cm at Cheticamp, NS, 1964–5 [Environment Canada 1982]; 648 cm at St John's, 2000–1 [National Climate Data and Weather Archive, www.climate.weatheroffice.ec.gc.ca]) occur throughout the region. All four provinces were affected by each of these types of natural events during the period 1989–2010.

Slope failures have resulted in more than 50 deaths on the island of Newfoundland since 1900 (Batterson et al. 1999; Liverman et al. 2001; Liverman, Batterson, and Taylor 2003; Liverman, Catto, and Batterson 2006) and have posed significant problems in New Brunswick and Nova Scotia. Coastal erosion in Prince Edward Island has caused similar problems. Coastal erosion has been ubiquitous in all areas of Atlantic Canada where unconsolidated sediment, driven by rising sea levels and glacio-isostatic subsidence, forms the shoreline (e.g., Forbes et al. 2004; O'Reilly, Forbes, and

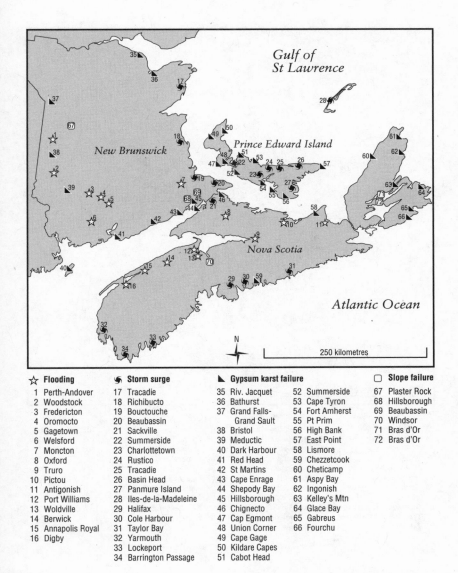

Figure 10.1a
Selected locations with natural hazards, New Brunswick, Nova Scotia,
and Prince Edward Island, Canada

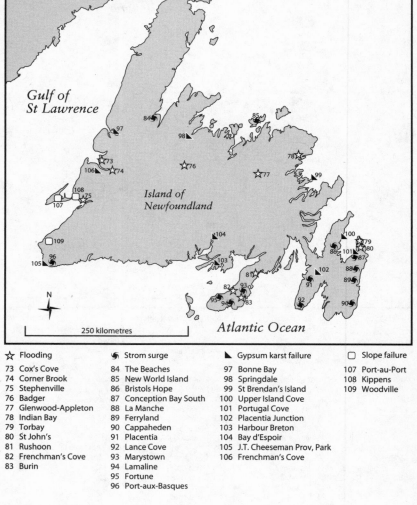

Figure 10.1b
Selected locations with natural hazards, Island of Newfoundland, Canada

☆ Flooding   🌀 Strom surge   ◣ Gypsum karst failure   ▢ Slope failure

73 Cox's Cove	84 The Beaches	97 Bonne Bay	107 Port-au-Port
74 Corner Brook	85 New World Island	98 Springdale	108 Kippens
75 Stephenville	86 Bristols Hope	99 St Brendan's Island	109 Woodville
76 Badger	87 Conception Bay South	100 Upper Island Cove	
77 Glenwood-Appleton	88 La Manche	101 Portugal Cove	
78 Indian Bay	89 Ferryland	102 Placentia Junction	
79 Torbay	90 Cappaheden	103 Harbour Breton	
80 St John's	91 Placentia	104 Bay d'Espoir	
81 Rushoon	92 Lance Cove	105 J.T. Cheeseman Prov, Park	
82 Frenchman's Cove	93 Marystown	106 Frenchman's Cove	
83 Burin	94 Lamaline		
	95 Fortune		
	96 Port-aux-Basques		

Parkes 2005; Webster et al. 2006; Vasseur and Catto 2008; Thompson, Bernier, and Chan 2009). Avalanches have occurred in New Brunswick and Nova Scotia and have killed at least 30 people in insular Newfoundland since 1833 (Liverman 2008). The region's earliest recorded earthquake, in 1764, affected Saint John, as reported in the *Halifax Gazette* (Burke 1984). Terrestrial-focus earthquakes as powerful as M5.7 have affected the Miramichi area periodically over the past 30 years (e.g., Stevens 1983), and the 28 people killed during the 1929 Burin Tsunami currently stands as the greatest recorded death toll from a single seismic event in Canadian history (Hyndman, Hyndman, and Catto 2008).

The perception that Atlantic Canada is relatively untouched by natural hazards thus does not accord with reality. The existence of this perception, however, has consequences in terms of the region's identification and assessment of, preparedness for, and effective response to natural hazards, both those commonly anticipated in the region (such as hurricanes and winter storms) and those that are less frequent (such as tsunamis). Identification, mapping, and vulnerability assessment thus pose a series of challenges in the region, although progress continues to be made. This chapter considers the identification, mapping, and vulnerability assessment in the Maritime provinces and insular Newfoundland. Labrador represents a separate set of challenges, due to its distinctive physical and human geography, and is thus worthy of separate treatment.

The effects of different natural hazards are frequently linked, accentuating risk, increasing vulnerability, and augmenting damage and human cost. Ongoing climate change could intensify many of these natural hazards (Vasseur and Catto 2008; Catto 2010). Hazards are frequently accentuated by human choices and disregard of known best practices, as exemplified by construction in floodplains, in exposed coastal sites, beneath unstable frost-shattered slopes, and on active gypsum karst terrain. The discussion here focuses on some examples of natural hazards in Atlantic Canada and is intended to illustrate the styles of investigation that have been conducted by the author up to November 2010.

EXAMPLES OF NATURAL HAZARD INVESTIGATIONS
IN ATLANTIC CANADA

Storms, Storm Surges, and Rising Sea Level

Atlantic Canada was subjected to significant storm activity during the period 1989–2010, producing heavy rainfall and storm surges that destroyed property in all four provinces. Hurricane Juan (2003), the most expensive hurricane in Atlantic Canadian history, killed eight people and caused at least $200 million in damage to Nova Scotia and Prince Edward Island (Environment Canada 2004). In addition to true hurricanes (such as Juan), many of the most damaging summer and early autumn events result from the interaction of northeastward-tracking hurricanes or tropical storms with mid-latitude storms approaching from the west, or low-pressure systems (e.g., Igor, which hit eastern Newfoundland in 2010; Figure 10.2). These "extratropical transitions" or "Canadian hurricanes" change direction and velocity quickly and carry heavy rainfall (Hyndman, Hyndman, and Catto 2008). Wind speeds are generally lower than those for hurricanes (maximum Saffir-Simpson values of ~2). As a result, most extratropical transitions affecting Atlantic Canada bring heavy rainfall rather than extreme winds and consequent storm surges.

Although true hurricanes affect southern Nova Scotia, the majority of summer and autumn storms hitting the region involve extratropical transitions. Significant events in the past 20 years have included Bob (1991), Luis (1995), Opal (1995), Hortense (1996), Floyd (1999), Gert-Harvey (1999), Irene (1999), Michael (2000), Gabrielle (2001), Gustav (2002), Juan (2003), Rita (2005), Florence (2006), Chantal (2007), Bill (2009), and Igor (2010) (see, e.g., Taylor et al. 1996a; 1996b; 1997; Parkes and Ketch 2002; Catto, Scruton, and Ollerhead 2003; Catto and Hickman 2004; Smith et al. 2004a; 2004b; O'Reilly, Forbes, and Parkes 2005; Catto 2006; 2007; Catto et al. 2006; Hickman 2006; Brake 2008; Vasseur and Catto 2008; Catto and Catto 2009; Catto and Tomblin 2009b; Thompson, Bernier, and Chan 2009; Catto 2010). Preliminary estimates suggest that Igor caused at least $100 million in damage to eastern Newfoundland in September 2010, as well as taking one human life.

The relationship between changes in frequency and magnitude of North Atlantic hurricanes and increases in air temperature or sea surface temperature is not clear at present, and consensus does not exist among hurricane specialists (for differing recent opinions, see Landsea et al. 2006; Kossin et al. 2007; Elsner, Kossin, and Jagger 2008; Emanuel, Sundararajan, and Williams 2008; Saunders and Lea 2008). Difficulties have centred on the link between general changes in climate and specific weather events and the lack of detailed historical records (e.g., Chang and Guo 2007). In addition, because of the variations in hurricane track patterns and the interaction to form extratropical transitions, the number of summer and early autumn storms in Maritime Canada is not linked to the total number of hurricanes occurring over the southwestern North

Figure 10.2
Hurricane Igor caused extensive flooding and washouts in the Burin and Bonavista peninsulas of Newfoundland in September 2010. Photo shows the centre of Trouty, Bonavista Peninsula. *Source*: Public Safety Canada.

Atlantic or Caribbean Sea in any particular year (Catto 2006; 2007; see Figure 10.3). Overall tropical storm frequency in the North Atlantic cannot be correlated with temperature variations in Atlantic Canada. Consequently, a cautionary policy would anticipate no significant changes in hurricane frequency for Atlantic Canada but also the continuing likelihood of significant events (Vasseur and Catto 2008).

Southwesterly and westerly winds associated with late autumn and early winter storms also are significant, as the coastlines of Atlantic Nova Scotia, the Bay of Fundy, and southern Newfoundland typically remain entirely ice-free until February at a minimum. During some years, these coastlines may remain ice-free throughout the winter, although most strong storms occur prior to February. Coastlines along the Gulf of St Lawrence and in northeastern Newfoundland generally are affected by offshore ice in winter, although freezing (with consequent exposure to storm action) occurs late during El Niño–influenced years (e.g., 1996–8). Extensive ice cover

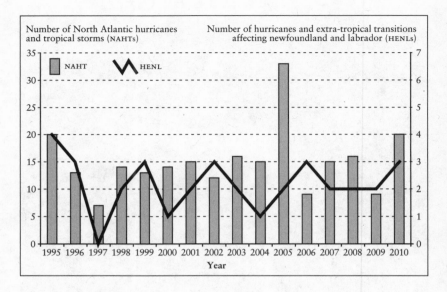

Figure 10.3
Comparison of the total number of North Atlantic hurricanes and tropical storms
(NAHTs) recorded by NOAA (National Oceanic and Atmospheric Administration,
USA) with the number of hurricanes and extratropical transitions impacting
Newfoundland and Labrador (HENLs), Canada, 1995–2010. Years of enhanced or
reduced tropical storm activity throughout the North Atlantic do not correspond
with variations in storm frequency (or intensity) in Newfoundland and Labrador.

in some years protects shorelines from winter storms in the southern
Gulf of St Lawrence and along the south coast of Newfoundland.

Storms, and consequent storm surges, are significant natural hazards during autumn, early winter, and spring (see, e.g., Taylor et al.
1996a; Forbes et al. 2000; 2004; McCulloch, Forbes, and Shaw
2002; Catto et al. 2003; 2006; Catto and Hickman 2004; Wright
2004; Catto 2006; Hickman 2006; Webster et al. 2006; Brake 2008;
Vasseur and Catto 2008; Catto and Catto 2009). Since 1990, storms
have caused significant damage at localities along the Gulf of
St Lawrence (e.g., Beaubassin and Bouctouche, NB; Basin Head and
Panmure Island, PEI) and along the exposed Atlantic coastlines of
Newfoundland (e.g., The Beaches, Ferryland, Lamaline, and New
World Island). Impact has involved repeated destruction and dam-
age to dwellings (e.g., Cap Pele, NB; Charlottetown, PEI; Yarmouth,
NS; Lamaline, NL), damage to infrastructure at key installations,
such as the ferry terminal at Channel-Port-aux-Basques and the

Maritime Electric facility at Charlottetown (January 2000), and effects on beaches and dunes, with consequent damage to tourist attractions (e.g., Bouctouche Spit and Parlee Beach, NB; Basin Head, Panmure Island, Rustico, and Tracadie, PEI; Cole Harbour, Cow Bay, and Taylor Bay, NS; J.T. Cheeseman Provincial Park and Lance Cove, NL).

Northeasterly storms are the most effective agents of erosion and property destruction for communities facing the open Atlantic from St Anthony to Cape Race. A severe northeasterly storm surge struck the village of La Manche in January 1966, fortunately with no casualties: it completely destroyed all flakes and fishing infrastructure, all the boats, the suspension bridge across the harbour, and most of the buildings, leaving only foundations. Following this event, residents abandoned the community. The same event caused destruction in all communities from Cappaheden north to The Battery (St John's), severely affecting fishery operations.

Individual shoreline segments respond to large-scale factors ranging from regional to hemispheric, including overall storm activity and variations in atmospheric pressure measured by the North Atlantic oscillation (NAO). The impact of any particular storm on an individual beach depends on the angle of wave attack, the number of previous events during the season, and other local factors. The effectiveness of any particular storm as a geomorphic agent or hazard does not depend on the overall number of storms during that year (Taylor et al. 1996; Catto, Scruton, and Ollerhead 2003; Catto et al. 2006).

Variations in the NAO are reflected in the amount of snow and ice cover on the beaches, sea ice activity, and northeasterly storm effectiveness during autumn, winter, and spring. Consequently, there is a general correlation between the NAO and winter storm erosion on beaches facing northeast, resulting in coarser beaches with steeper profiles (Catto 2006). Local factors, however, determine the results at any particular beach. Observations throughout the period 1989–2010 in several locations in Atlantic Canada indicate that no single beach can serve effectively as a proxy to assess the overall impact of climate variation (Vasseur and Catto 2008; Catto 2010).

The hazards resulting from storm activity are increasing in Atlantic Canada as a result of ongoing sea level rise (Taylor et al. 1996; Shaw et al. 1998; 2001; Hilmi et al. 2002; Vasseur and Catto 2008). Archaeological sites at Ferryland, NL (Catto, Forbes, and Liverman

2000; Catto, Scruton, and Ollerhead 2003), Fort Beausejour, NB (Scott and Greenberg 1983; Shaw and Ceman 1999), Louisbourg, NS (Taylor et al. 2000), and St Peter's Bay, PEI (Josenhans and Lehman 1999; Shaw, Gareau, and Courtney 2002), among others, indicate that the sea level has risen since about 1600. Evidence of enhanced erosion along many Atlantic Canadian beaches, and inundation of terrestrial peat deposits and trees, suggest that transgression is currently occurring. Coastal erosion accelerated by rising sea levels has taken place at several localities, notably along the south coast of Nova Scotia (Taylor et al. 1985; 1996b; Shaw, Taylor, and Forbes 1993; Shaw et al. 1994), in eastern New Brunswick (Ollerhead and Davidson-Arnott 1995; Shaw et al. 1998; Vasseur and Catto 2008), along all coasts of Prince Edward Island (Forbes et al. 2002; McCulloch, Forbes, and Shaw 2002; Vasseur and Catto 2008; Catto and Catto 2009), and along Conception Bay in Newfoundland (Taylor 1994; Batterson et al. 1999; Catto, Scruton, and Ollerhead 2003; Smith et al. 2004a; 2004b; Batterson, McCuaig, and Taylor 2006).

The rates of submergence, estimated from submerged ^{14}C-dated terrestrial deposits and from tide gauge records, are 1–5 mm/year across Atlantic Canada (see Shaw et al. 1998; Catto, Forbes, and Liverman 2000; Vasseur and Catto 2008). Exact quantification of rates of sea level rise over short periods (ranging from decades to hundreds of years) is complicated by many factors, including local subsidence (cf. Belpeiro 1993), confusion of storm and tsunami deposits with those associated with modal marine conditions (Dawson 1999), erosion induced above mean high water (e.g., Bryan and Stephens 1993), and landward migration of coastal gravel barrier beaches (Taylor et al. 1985; Forbes et al. 1995; Orford et al. 1995; Orford, Carter, and Jennings 1996).

Consideration of the geological factors, rate of sea level rise, amount of coastal erosion, wave climate, and tidal regime allows calculation of the sensitivity to sea level rise of shoreline segments (e.g., Gornitz 1990; 1991; 1993; Gornitz, White, and Cushman 1991; Gornitz et al. 1993). This assessment has been completed for Atlantic Canada as a whole on a broad regional scale (Shaw et al. 1998), and more detailed assessments have been conducted for specific segments of the coastline (e.g., Catto, Scruton, and Ollerhead 2003).

Although a long-term erosion rate is a useful guide to the establishment of set-back limits (Taylor 1994) and indicates where specific structures are in danger, it does not fully indicate the true hazard potential at a particular site. As most of the erosion is accomplished by individual storms, hazard assessment requires consideration of the probability of the maximum impact of a particular storm, rather than monitoring and dealing with incremental, infinitesimal removal of sediment on a daily basis. The Point Verde lighthouse site (Placentia) has the longest (semi-quantitative) record of cliff erosion assessment in eastern Newfoundland, but it does not include potentially major events such as the "Great Independence Hurricane" of 1775 (Stevens and Staveley 1991; Ruffman 1995b; 1996; Stevens 1995) or the tsunamis of 1864 (suspected) and 1929 (Ruffman 1995a; 1995b). These events, or future occurrences of a similar magnitude, have the potential to cause much more erosion. The 1775 hurricane and the 1929 Burin Tsunami caused coastal erosion and damage to structures in localities that are not generally subject to high-energy events. Monitoring at other sites does not extend back before the initial observations of Forbes (1984). The absence of long-term records means that present erosion rates may not indicate the magnitude of previous (or future) events.

Flooding

Floods in Atlantic Canadian communities involve combinations of one (or more) natural causes directly related to climate coupled with anthropogenic factors. Hurricanes, extratropical transitions, and autumn and winter storms generate both river flooding and coastal storm surges. Storm-related flooding is a significant problem in Newfoundland on the Avalon Peninsula (e.g., during extratropical storm Gabrielle in 2001) and on the Bonavista and Burin peninsulas (e.g., during Igor, September 2010), along Nova Scotia's Atlantic coast (November 2010), and in Prince Edward Island. Autumn and winter storms create coastal flooding hazards along the Gulf of St Lawrence coastlines of New Brunswick, Prince Edward Island, and Newfoundland; the Northumberland Strait coastlines of Nova Scotia and Prince Edward Island; and the coastlines of northern, eastern, and southern Newfoundland. Short fluvial systems are more susceptible to storm-related flooding than are longer systems, as the response time between precipitation and flooding is short. Precipitation is rapidly distributed to rivers and streams, triggering flooding within an hour after the initiation of heavy rainfall (see Watt 1989; Hirschboeck, Ely, and Maddox 2000; Wohl 2000; Catto and Hickman 2004; Hickman 2006; Brake 2008).

Rain-on-snow events occur in late winter or early spring in all areas of Atlantic Canada. In southerly areas subject to midwinter thaws associated with warm southwesterly winds, they are also involved in some winter events. Recently affected areas include Corner Brook, NL, Gagetown, NB, and Oxford, NS.

Rain-on-snow events are common because of the number of temperature fluctuations around the freezing point. Initially, rain falling on snow tends to run off the surface, particularly if the surface of the snowpack has previously partially melted and refrozen, producing a hard crust of impermeable ice crystals. The same effect results from repeated freezing rain, which can build up a hard crust of adhering ice on the surface of the snowpack.

Where rain is unable to penetrate an ice-encrusted snowpack, runoff is rapid, resulting in flooding (Quick and Pipes 1976; Watt 1989). Frequent freezing rain thus reduces rain-on-snow flooding. If rain persists, or if the snowpack is not ice-encrusted, the entire snowpack may become mobilized, contributing the snow to the river systems. Water-saturated failures of snow banks are most likely where the snowpack is thickest, on the steepest slopes.

Rain-on-snow failures pose an additional problem, due to their season. If the flood event is followed by a colder episode, the floodwaters may freeze in position, hampering efforts at re-establishing infrastructure and rehabilitating property, and perhaps increasing property damage through ice pressure. Refreezing may also interrupt sewage discharge. Refreezing of floodwaters was a significant problem in the February 2003 event at Badger, NL (Peddle 2004; Figure 10.4), and has been observed in other floods, including those at Rushoon, NL (Canada–Newfoundland Flood Damage Reduction Program, 1990). Rapid snowmelt also can trigger flooding. In Newfoundland, flooding in the Indian Bay (Bonavista Bay) and Glenwood-Appleton area (Gander River) during spring 2004 resulted from rapid snowmelt, causing the flooding of a cemetery and temporary evacuation of two homes.

Ice jamming is responsible for repeated flooding along the Saint John River through New Brunswick (e.g., Gagetown, Perth-Andover, and Woodstock; see Ismail and Davis 1992; Hare, Dickison, and Ismail 1997; Beltaos 1999; Beltaos, Ismail, and

Figure 10.4
Flooding resulting from ice jamming, Badger, Newfoundland, February 2003
(photo: D. Hawco)

Burrell 2003) and at Badger on the Exploits River in Newfoundland
in February 2003 (Peddle 2004). Ice-jam floods in both rivers can
result from either thermal effects, such as rapid changes of the water
surface temperature, commonly during freeze-up, or dynamic causes, such as thrusting of drifting ice against obstructions or at con-
strictions in the channel (Pariset, Hausser, and Gagnon 1966; Beltaos
1983; Watt 1989; Beltaos and Burrell 2002; Beltaos, Ismail, and
Burrell 2003). In Atlantic Canada, dynamic events appear more
common, although ongoing climate warming could result in more
thermal events. Areas subject to repeated ice jams are marked by
disruptions to the vegetation, including broken and ice-scarred trees,
which can be readily recognized through field investigations (e.g.,
Pennell 1993; Nichols 1995).

Ice jamming is often associated with the accumulation of debris in
a river system, through either natural or artificial causes. During a
highly dynamic ice jam, the moving ice erodes the banks, ripping
away trees, large masses of frozen ground, fractured bedrock, and

artificial materials. This debris often is thrust against downstream banks or grounds on the stream bottom, contributing to the formation of the jam. Ice-jam debris within communities, however, is commonly removed by residents, to prevent further flooding, remove obstructions to river use, improve the aesthetics of the stream, or remove brush to deny habitat to undesirable insects. Consequently, the distribution of ice jam–transported brush is not a reliable guide to the extent of ice jamming.

Coastal ice jams result from wind-driven accumulations of offshore ice at river mouths and lagoon outlets. They form where shallow, narrow rivers or tidal channels are confined laterally by gravel bars, which face the direction of the prevailing or strongest winds. Formation is also governed by prevailing currents and by the availability of both sea ice and freshwater ice transported to the sea from rivers. Coastal ice jams can form in some outlets where there is no marine ice present, but where quantities of river or lagoonal ice can be driven back to the shore by strong winds.

Although the assessment of the flood hazard usually involves mapping of the flood plain and determination of previous flood extents, excessive precipitation is responsible for other hazards as well. The inter-relationship between precipitation, flooding, and slope failures documented for the Corner Brook–Humber Arm area (Catto and Hickman 2004; see also Batterson, McCuaig, and Taylor 2006) indicates the value of combined multi-hazard assessment. Hurricanes that generate precipitation simultaneously generate storm surges throughout Atlantic Canada. Examination of all natural hazards would allow the construction of maps illustrating the vulnerability of communities to these phenomena, many of which are linked. At present, this approach has not been employed to investigate any community in the region (Catto and Tomblin 2009a; 2009b). Such a comprehensive examination would be in many ways more effective and significant than focusing solely on a single "hazard."

Slope Failures

Gravitational slope failures conventionally are not thought to be a major hazard in Atlantic Canada (e.g., Evans 2001). The list of Canadian landslide disasters (three or more fatalities) compiled by Evans (1999; 2001) cites only two events in the region. It is misleading to assume, however, that landslides are insignificant in the area.

Batterson et al. (1999), Liverman et al. (2001), and Liverman, Batterson, and Taylor (2003) reviewed geological hazards in insular Newfoundland and found more than 50 fatalities from 18 slope failures since 1823, as well as at least 30 from snow avalanches. Although no deaths are known from the Maritime provinces, slope failures and mass movements are common and affect transportation, forestry, coastal and urban development, and other human activities.

Rock fall is common in Atlantic Canada, because of the combination of relief, steep slopes resulting from glaciation, repetitive freeze-thaw conditions, and jointed, faulted, and stratified rock units with multiple planes of weakness. Most incidents involve single blocks, but they do cause damage, commonly affecting highways or resulting in coastal erosion. Investigations generally occur after an incident, rather than involving prior mapping or anticipatory zoning and municipal planning.

In Newfoundland, examples include several small failures along Pitts Memorial Drive in St John's between 1990 and 2009; rock fall and failures of bedding plane slopes relating to highway construction around Bay d'Espoir, Bonne Bay, Placentia Junction, and Port-aux-Basques; rock falls onto residential property in Portugal Cove, Springdale, and Upper Island Cove; and repetitive rock falls at natural and artificial exposures on St Brendan's Island, Bonavista Bay (White 2002).

Nova Scotia also suffers from frequent slope failures, notably in Cape Breton Island, but there is no record of fatalities. Numerous scars from debris torrents or avalanches have been identified in the Cape Breton Highlands (Finck 1993; Grant 1994). Large landslides appear to have taken place along the Aspy Scarp and near the mouth of the Cheticamp River (Grant 1994), although there is limited documentation. In terms of impact on people, a series of slope failures almost completely destroyed part of the Trans-Canada Highway across Kelly's Mountain in 1982.

Along the Gulf of St Lawrence and Northumberland Strait coastlines, the most common form of failure involves erosion of weak Carboniferous and Permian sandstones and shales. Overlying glacial deposits are commonly involved, especially on Prince Edward Island. Incremental slope failures result from frost action coupled with marine undercutting induced by rising sea level, about 3 mm/year throughout the region (Shaw et al. 1998; Forbes et al. 2002; McCulloch, Forbes, and Shaw 2002; Vasseur and Catto 2008).

Locally, saturation of the bedrock and overlying sediments resulting from agricultural practices (irrigation, disruption of drainage) has contributed to debris flow and creep failures, but anthropogenic triggers are nowhere of primary importance.

Although erosion is generally marked by incremental block failure, making generalization from the calculation of short-term rates suspect, rates for bedrock-seated failures have exceeded 80 cm/year at several locations in Prince Edward Island, including Cape Gage, Cape Kildare (Genest and Joseph 1989), and Panmure Island. Around the island province, the erosion of bedrock-supported cliffs at rates of about 10–50 cm/year is also evident at East Point (John Shaw, Geological Survey of Canada, personal communication), High Bank, Point Prim, Fort Amherst National Historic Site, MacCallums Point, Cap Egmont, Cabot Beach Provincial Park (Catto 1998; Catto and Catto 2009), Cape Tryon, and Union Corner (Figure 10.5). Monitoring of coastal cliffs is relatively limited, however, consisting largely of spot observations not incorporated into a systematic effort, and coastal erosion has not been quantified rigorously, with observations not reported from many localities. The potential scale and extent of coastal erosion of bedrock shores in Prince Edward Island, as well as along the Acadian Peninsula of New Brunswick, indicate that much more extensive research is desirable.

Along New Brunswick's Baie des Chaleurs west of Bathurst, Palaeozoic limestone/shale successions crop out. These units are subject to local block failure triggered by frost action, particularly along the Restigouche Estuary. Karst erosion and biochemical dissolution from seabird droppings also contribute to joint widening and eventual failure by heaving and toppling. Susceptibility is controlled primarily by joint orientation and the degree of bird activity and is related inversely to vegetation cover. The aspect with respect to the prevailing westerly winds plays a much lesser role. Rates of erosion appear to be low – about 10 cm/year or less – although little quantitative analysis has been conducted. Failure in interior sites is associated with quarrying, mining, and road cuts, involving pre-existing joints accentuated by glacio-isostatic recovery that are reactivated or lubricated by heavy rainfall (e.g., Park and Broster 1996).

In the vicinity of Saint John, slope failures involve glacial sediment at Lorneville Cove (Ruitenberg, McCutcheon, and Vanugopal 1976; Ruitenberg and McCutcheon 1982), marine and glacial sediment and the underlying Triassic sedimentary rock (Tanoli and Pickerill 1988) at Red Head and vicinity, and glacial sediment and the

Figure 10.5
Coastal erosion at Union Corner Provincial Park, Prince Edward Island

underlying Carboniferous sedimentary rock at Mispec Point and western Mispec Bay. The weak sedimentary Carboniferous and Triassic rocks are susceptible to coastal erosion. At Red Head, where activity has been most consistent throughout the past 15 years, coastal erosion in association with sea level rise, hurricanes, and heavy precipitation has resulted in a series of debris flows. Overloading of the bluffs due to suburban development and saturation of the deposits have enhanced susceptibility to slope failure. Removal of vegetation has facilitated frost penetration on southwest-facing bluffs (facing the prevailing wind), resulting in destabilization and initiation of creep, in turn promoting the development of shallow-seated debris flows, local movement of frozen slabs of sediment, and minor slumping. Residents' efforts at stabilization have focused on protection at the waterline, involving the installation of chained mats of automobile tires and gabions, but have proven ineffective. In addition to other localities in the Saint John area, similar minor failures involving glacial sediment and underlying friable bedrock occur at Cape Enrage (Chignecto Bay), Cape Maringouin (Shepody Bay), and Mares Bay.

The interiors of Prince Edward Island and most of New Brunswick are not subject to significant slope failures because of their limited topographic relief, with no notable events recorded in Fredericton (cf. Broster 1998) or Moncton. Local undercutting of cut banks along the upper reaches of the Saint John River has resulted in minor rotational slides and block failures, particularly where clays overlie glacial deposits. Failures are also evident along the east bank of the Saint John River, south of Grand Falls–Grand Sault and at Bristol, and along the west bank at Meductic.

Dissolution of Gypsum Karst

Karst topography results from chemical dissolution of carbonate limestone and dolomite or chemically precipitated evaporate rocks, including halite, sylvite (potash), and gypsum. Areas of exposed gypsum in Atlantic Canada, including Plaster Rock and Westmorland County, NB (e.g., Webb 2002); Windsor and the north shore of the Bras d'Or Lakes, NS (e.g., Adams 1993); and the Port-au-Port Peninsula, Stephenville, and Woodville, NL (e.g., Knight 1983; House and Catto 2008; Figure 10.6), are subject to these dissolution processes.

Gypsum, deposited as a chemical precipitate, is consolidated to form geologic beds, but it is easily deformed during folding and faulting. When the gypsum beds are exposed to water that is not saturated with either calcium or sulphate, dissolution occurs. The rate of dissolution depends on the concentration of sulphate ions in the water (low concentrations promoting dissolution), the concentration of hydrogen ions (high concentrations or acidic water favouring dissolution), and the presence of humic acids and organic compounds in the water. In addition to these chemical factors, the volume of water flowing through or across the gypsum surface (discharge), the duration of contact between individual water molecules and gypsum crystals, and the turbulence of the water also influence the rate of dissolution. Temperature is also a factor, but the relatively low temperatures common in Atlantic Canada limit its importance for gypsum dissolution, as optimal conditions for dissolving the rock require relative warmth.

Under normal circumstances, the rate of surface dissolution over a flat expanse of gypsum would be on the order of mm/100 years. However, dissolution speeds up when the gypsum beds are confined laterally by other rock units that are not susceptible to dissolution, where dissolution is concentrated locally by wave action (as on

Figure 10.6
Gypsum karst doline, Woodville, Newfoundland (photo: K. House)

coastal cliffs) or by streams, or where dissolution occurs beneath the surface. As well, increases in precipitation through climate change, or in the rate at which precipitation enters the groundwater system, resulting from clearing of forest cover, will also increase dissolution. All these factors are involved in Atlantic Canada (House and Catto 2008).

The dominant landforms are suffusion dolines, from circular to oval and in plain view, with some of the largest exceeding 50 m in diameter (House and Catto 2008). Walls are generally steep, with modal slopes typically 60° at the surface, increasing to nearly vertical at depth. Water levels at the bottoms of the dolines vary significantly, depending both on recent precipitation and on drainage through the bases. Differing water levels noted in adjacent dolines indicate that not all features are interconnected to a common water table.

The development of the suffusion dolines may not be evident on the surface. At Hillsborough, Port-au-Port, Windsor, and Woodville, the gypsum beds are overlain by glacial sediment. The glacial deposits do not contain large quantities of sediment derived from the underlying gypsum. Consequently, they are not subject to dissolution but instead form blankets on the surface, concealing the underlying gypsum units. As the underlying gypsum is dissolved, the overlying glacial sediment may be sufficiently thick or rigid to permit formation of cavities, which can be detected with ground-penetrating radar. In other locations, progressive dissolution of the underlying gypsum causes the surface blanket to gradually sag into the developing low area.

Failure, producing a sinkhole evident on the surface, occurs when the amount of gypsum dissolved is sufficient to cause the blanket to fail. This results in the collapse of the glacial sediment blanket into the doline, partially filling the bottom with disturbed material. Dissolution of the gypsum will continue, however, as water infiltrates around the glacial materials.

Attempts to infill the dolines by adding sediment, vehicles, and refuse have been unsuccessful, as dissolution will continue as long as either surface or ground water is in contact with the gypsum (House and Catto 2008). The addition of refuse containing acids or other chemical compounds may actually accelerate gypsum dissolution, by changing the chemistry of the water in contact with the gypsum beds. As well, the addition of waste into the sinkholes may pollute the groundwater system, thus posing possible danger to residents.

At Woodville, mapping has revealed a minimum of 70 dolines in differing stages of development, ranging from hints of depressions in a field to spectacular holes filled with water (House and Catto 2008). Most appear to be actively propagating and expanding, and several oval dolines have extended laterally to join together to form linear features. Properties adjacent to the doline walls will eventually suffer collapse as dissolution continues. Study is ongoing to determine if the rates of dissolution and doline development have increased over the past 50 years and to assess whether removal of ground cover by farming or logging has speeded karst dissolution. The removal of vegetation would allow more water to infiltrate the topsoil, speeding gypsum dissolution.

CHALLENGES IN HAZARD ASSESSMENT

The foregoing list of natural hazards for Atlantic Canada is impressive and diverse. With its tradition of strong community spirit and interest in local matters, combined with general acceptance of an important role for provincial and federal governments, the region would be expected to have programs for detailed hazard mapping and assessment. However, this is not the case, and the identification of natural hazards, assessment of their impact, and suggestions for adaptation have proceeded in somewhat piecemeal fashion.

The studies we discussed above are typical of those undertaken to assess all types of natural hazards in Atlantic Canada. Most have focused on the identification of a particular hazard in a specific area and have not investigated all hazards within a community. Most also have been conducted from primarily a geological or a geomorphological viewpoint, with limited or no input concerning socioeconomic aspects. Although some recent studies have considered the concepts of exposure and vulnerability implicitly or explicitly (see Catto, Forbes, and Liverman 2000; Catto, Scruton, and Ollerhead 2003; Catto et al. 2006; Catto 2008; Catto and Tomblin, in press), most have focused on identifying sensitive areas or conditions conducive to hazardous events.

Although mapping of some hazards is available for some localities, comprehensive regional mapping and assessment of any one type of natural hazard is not yet available. A number of factors are responsible: fragmentation of databases; loss of records and of institutional, community, and individual memories; focus on a single hazard or agency; limited financial and personnel resources; and difficulties with effective dissemination of the available information (Catto and Tomblin 2009a; 2009b; in press).

The fragmentation of databases has affected the study of all types of natural hazards in the region. Efforts under way by the Geological Survey of Canada through the Canada Landslide Project (P.T. Bobrowsky, personal communication; sts.gsc.nrcan.gc.ca/landslides/index_e.asp) to reconstitute the record of slope failures throughout the Atlantic provinces (see Evans 1999; 2001; Liverman et al. 2001; Liverman, Batterson, and Taylor 2003) show how many data have disappeared. Many events were recorded only in the oral histories of communities. Government records have proven to be misplaced or absent. Efforts to trace the records of the Canadian National Railways' Newfoundland division, which could have identified slope failures and weather-related delays, have been unsuccessful (e.g., Catto et al. 2006). Similar difficulties have become apparent in attempts to investigate flooding history (e.g., Catto and Hickman 2004; Hickman 2006; Liverman, Catto, and Batterson 2006; Brake 2008).

The failures of institutional and community memory have led to the underestimation of natural hazards, particularly for events with low recurrence frequencies. An example is provided by Deadman's Gulch, at Ferryland, NL (Liverman, Batterson, and Taylor 2003). In the initial published report, J.W. White (1902, 19–20) stated that this event occurred in 1823 "or thereabouts." White provided a detailed account of the deaths of 42 fishermen, who took shelter from a storm in a sea cave and were killed when the cave roof collapsed. The bodies were not recovered, "having been buried, no doubt, beneath the crushing mass of stone which had fallen on them." If authentic, this would be the worst geologically related disaster in Newfoundland and the third-worst disaster resulting from gravitational slope failure in Canadian history (cf. Evans 1999). However, corroboration of this story has proven difficult (Liverman et al. 2003), as there are no extant newspaper accounts or government records. Investigation of parish records has also failed to show definitive evidence, and the incident has not been recorded in the oral histories of Ferryland residents. Physical evidence on the surface indicates that an elongate depression, created by collapse from below, extends from the cliff edge inland approximately 200 m, but no method of dating the time of formation of the depression currently exists. At present, therefore, the evidence for the deaths at Deadman's Gulch rests almost solely on the account of J.W. White (1902), written more than 80 years after the event. If it occurred, the disaster has passed from the collective memory of Ferryland.

A more recent example is provided by Harbour Breton, on the south coast of Newfoundland. On 1 August 1973, following several weeks of heavy rainfall, a landslide swept four houses into the harbour, killing four children. Fifteen families were subsequently resettled, leaving the affected area and the flanks of the landslide scar vacant. The area in which the event occurred had been the site of a previous landslide, about 1953, based on the identification of a landslide scar on aerial photographs. At that time, however, there were no houses in the area, and consequently there was no incident report filed on which to base future planning.

On at least three subsequent occasions, in November 1982, February 1984, and June 1986, provincial geologists visited Harbour Breton to examine potential rock fall and landslide problems, an indication of heightened awareness of the hazard in the concerned institution (Liverman et al. 2001). Although the slope failure zone remained apparent on the landscape, applications to build new houses in the affected and now vacant zone have been received by the town of Harbour Breton and were rejected only after comments from the Ministry of Natural Resources. The potential for future landslides had not been entirely recognized by many residents.

Memories within communities have proven to be relatively short in many instances. In combination with lack of understanding of the causes of previous events, and the likelihood of recurrence, this has led to the underestimation of problems. Examples include the lack of a flood response plan for Badger, NL (Peddle 2004), and the lack of perception of hurricane risk in Halifax prior to Hurricane Juan (Conrad and Hanson 2004). Following Juan's impact, statements that it was the worst hurricane in 100 years, coupled with the destruction of mature trees of similar longevity in Point Pleasant Park, led some residents to conclude that another similar hurricane will not strike the area for the foreseeable future (Bob Taylor and Don Forbes, Geological Survey of Canada, personal communications). Similar perceptions resulted from Environment Canada's unofficial assessment that flooding during Hurricane Igor in eastern Newfoundland represented a "100-year" event. The lack of understanding of the probabilities or recurrence frequencies of natural events, coupled with difficulties for researchers in expressing concepts of exposure (or "risk"), as we discuss below, lead to increased vulnerability.

Once natural hazards are assessed and sensitive areas mapped, the information must be effectively disseminated. Difficulties include establishment of credibility by researchers, particularly those outside the government, and determination of the most effective media and modes of communication. The effectiveness of efforts varies widely with the nature of the particular hazard and from community to community.

The entire coastline of the Maritime provinces and insular Newfoundland has been mapped by several government agencies, with a view towards the assessment of hazards resulting from hurricanes, storm surges, coastal erosion, sea level rise, and climate change. The information is available in government publications, in electronic form, and through consultation with individual researchers. Despite the wealth of material available, however, difficulties in dissemination still exist. Substantial differences exist between political jurisdictions, relating in part to attitudes and willingness to receive information expressed by political leaders, policy-makers, and residents.

Interest in coastal erosion by the government of Prince Edward Island led it to establish the Special Committee on Climate Change in February 2005. Community demands for information about coastal erosion at Basin Head, Panmure Island, PEI National Park, and other sites led to rapid dissemination of information and mapping from the Geological Survey of Canada. The active cooperation of the city of Charlottetown in studies of the effects of future storm surges, commissioned after the January 2000 event (Bruce 2002; McCulloch, Forbes, and Shaw 2002), is an excellent example of the effective dissemination of information and mapping of a natural hazard.

In New Brunswick, consultations on implementing the province's Coastal Management Plan have disseminated information about coastal erosion and sea level rise. Community involvement and demands for information have been prevalent. Coastal erosion along the southern Gulf of St Lawrence spurred establishment of the Southern Gulf of St Lawrence Coalition on Sustainability (SGCS) in 2004. Although the coastline had been mapped on several occasions by provincial and federal agencies, the SGCS pushed for effective dissemination of the information and for analysis of the future impact of coastal erosion and climate change on the economies and health of the communities. While the SGCS was initially driven by academic researchers who had been involved in hazard identification and mapping, its leadership has devolved largely to community groups and individuals concerned about local and regional issues. The SGCS sponsored a symposium in Bouctouche, NB, in November 2004 focusing on climate change and natural hazards in a regional context, rather than as isolated entities (SGCS 2004). In this instance, a community-based organization has taken the initiative in obtaining natural hazards information, rather than relying on government or academe. A key element in the SGCS's success: it engaged municipal, provincial, and federal politicians as coalition partners, rather than as targets for pressure.

Yet some crucial information in Atlantic Canada about coastal erosion, as well as other hazards (Catto, Forbes, and Liverman 2000; Catto and Tomblin 2009a; 2009b; in press), has not reached relevant parties. Despite mapping that recognized coastal erosion rates of 20–50 cm/year in Conception Bay South, NL, including reports identifying locations particularly sensitive to coastal erosion and sea level rise submitted to the municipality (Liverman, Forbes, and Boger 1994a; 1994b; T. Taylor 1994; Catto and St Croix 1998; Paone 2003; Pittman 2004; Smith et al. 2004a; 2004b), housing development of the Burnt Island, Granville Meadows, and Lions Club sites proceeded.

In contrast, effective communication of slope stability hazards along the South Side Hills in adjacent St John's led to city enforcement of a new requirement for geotechnical surveying prior to housing construction. Although the municipality was not fully aware of the previous history of slope failure, the information was received and acted on rapidly.

Current and previous efforts in Atlantic Canada have concentrated largely on the identification of hazards and assessment of local sensitivity to processes, rather than representing a true vulnerability assessment. In part, this reflects the regional gaps in basic mapping and hazard identification. An additional factor is the nature of financial and organizational support for natural hazards research, with many projects funded or administered by agencies interested primarily in the natural sciences. Until recently, most of the researchers were natural scientists, with limited grounding in social sciences, economics, or humanities.

The lack of connection between disciplines and agencies has led to duplication of effort. The mapping of coastal sediments, designed partly to assess hazards associated with storm waves, coastal erosion, sea level rise, and contamination by petroleum, was conducted in northeastern and eastern Newfoundland for Environment Canada (1988), Fisheries and Oceans Canada (e.g., Connors and Tuck 1999; see also O'Brien et al. 1998), and Natural Resources Canada (e.g., Edwardson et al. 1993, Owens 1994; Sherin and Edwardson 1996). The assessments were biophysical, despite the advantages that a multifaceted classification scheme would bring to the analysis of coastal hazards (cf. Cendrero and Fischer 1997; Cooper and McLaughlin 1998).

Identification and assessment of natural hazards have proceeded on a "single-hazard" basis, with studies focusing on the particular hazard (e.g., flooding, coastal erosion) rather than on the community. Multi-hazard approaches are uncommon and may be deemed to fall outside the mandate of the study. The incorporation of slope stability assessments into a study focusing on flooding, or incorporation of flooding created by overflowing rivers with that generated by storm surges in the same study of coastal communities (Catto and

Hickman 2004; Hickman 2006; Brake 2008), are the exceptions rather than the rule. The division of responsibility between government agencies, with some natural hazards classed as "environmental," others as "water-related," and others as "geological," also leads to fragmentation of effort. The assignment of flood mapping may rest with the government department charged with the environment, municipal planning, natural resources, or public services or may go to different ministries over time. The multiplicity of federal, provincial, and municipal agencies involved in various aspects of the coastal zone makes integrated, multi-hazard assessment a significant challenge (Catto and Tomblin, in press). As a consequence, comprehensive mapping and assessment of all natural hazards are not available for any community in Atlantic Canada.

ASSESSMENT OF EXPOSURE AND VULNERABILITY

The assessment of exposure, as distinct from sensitivity, has also been limited. In some circumstances, government agencies and university investigators have been reluctant to specify exposure quantitatively. Concerns expressed by informants, community leaders, and private citizens about decreases in property values or possible legal issues also constrain the assignment of quantitative measures of exposure.

Many hazard identification and mapping assessments in Atlantic Canada have been conducted primarily by geographers and geologists. Earth science training has traditionally focused on the occurrence of slope failures, coastal erosion, storm surges, earthquakes, or flooding "eventually," during geological time. Uncertainty is much greater about when a geological event will occur than about whether it will occur. Calculated rates of erosion commonly involve relatively short periods from a geological perspective (from a year to a decade). The available data sets for coastal erosion at Atlantic Canadian locations indicate substantial variation in rates over very short periods, with associated uncertainty in rates commonly in excess of 100 per cent (e.g., Genest and Joseph 1989; Liverman, Forbes, and Boger 1994a; 1994b; Catto, Scruton, and Ollerhead 2003).

Qualitative statements are preferred to semi-quantitative estimates under these circumstances. Unfortunately, such statements commonly appear imprecise, vague, and unreliable to people who do not have an earth science perspective, but who are assessing or managing risk and vulnerability (see Catto and Parewick 2008; Hyndman, Hyndman, and Catto 2008).

Human choices, such as construction on a flood plain, are seemingly not comprehensible from a strictly earth science or physical geographical viewpoint. From this perspective, *any* structure built on a flood plain may be considered to be "at risk." Assessment of a quantitative degree of exposure is complicated by the uncertainties surrounding the exact nature and magnitude of future climate change. As many natural hazards are triggered or accentuated by weather events (e.g., storms) rather than by overall long-term climate change, the quantitative assessment of exposure appears impractical and ultimately impossible. Unfortunately, this reality can lead to reluctance to make qualitative or relative statements concerning risk, leaving administrators, clients, and communities without guidance or input (Catto and Parewick 2008). Unless the natural hazard is linked to the consequences for a community in a tangible form, identification and mapping may be largely pointless, failing to influence the vulnerable community.

A successful approach, directly linking sensitivity to sea level rise with consequences for specific infrastructure, was demonstrated for Charlottetown by McCulloch et al. (2002; also see Bruce 2002). Cooperation from municipal authorities, coupled with a visually effective presentation incorporating detailed LIDAR analysis and computer graphics, led to rapid acceptance of the scientific findings and their socioeconomic implications, allowing the importance of sea level rise and storm surge activity to be widely recognized by the community.

Vulnerability assessment combines sensitivity and exposure analyses (Catto and Parewick 2008), and lack of detailed information about exposure prevents its realization for much of Atlantic Canada. Assessment of vulnerability integrates natural sciences, social sciences, humanities, and economics. Although true integration and collaboration among these groups are often called for and recognized as essential by all parties, particularly in workshop settings, they have proven "easier said than done." In Atlantic Canada, the primary reason appears to be lack of time for calm contemplation and in-field collaboration focusing on specific communities or suites of hazards. Institutional and organizational barriers, and lack of understanding, communication, and/or acceptance among different academic and professional disciplines, are also potential factors.

However, in Atlantic Canada, the professional, governmental, and academic communities focusing on natural hazards are relatively small. Issues of organization and discipline-specificity are less significant than are limited personnel and financial resources for on-site collaboration. Natural hazards investigators enjoined to "think outside their administrative or academic discipline box" are constrained by the demands of their inboxes of activities. Workshops and seminars, though valuable first steps, cannot be the total solution leading to multi-hazard vulnerability assessment: they must be followed by on-site collaboration among everyone working on actual hazard identification and vulnerability assessments.

The current and ever-improving state of GIS analysis should enhance the ability to initiate a multi-hazard approach to mapping and vulnerability assessment. Integration of layers of hazard mapping can allow common factors to be identified and relationships among natural hazards (and anthropogenic activities) to be recognized. Incorporation of geomorphological and sedimentological coastal mapping, assessment of sensitivity to sea level rise, and sensitivity to petroleum contamination into a single GIS system (Catto, Scruton, and Ollerhead 2003) allow comparison of these databases and serve as a base for further layers of coastal mapping.

However, data that are not effectively disseminated cannot be integrated; nor can data inadequately incorporated into metadata systems, or lost from archives (or within archives), or faded from memory. Maintenance of databases and metadata systems has frequently been accorded lower priority than collection of "new" natural hazard data, some of which will suffer the same fate. Ideally, from the beginning, the natural sciences–social sciences–humanities–economics partnership should also include data and metadata managers, to minimize problems with spatial representation and representativeness, quality control, and database compatibility. Although individual collaborations have proven successful in Atlantic Canada, a multi-hazard approach involving all relevant expertise has not been achieved.

CONCLUSION: FUTURE CHALLENGES

An integrated approach, encompassing exposure and vulnerability assessment of all natural hazards in concert, is a vital aspect of mitigating hazard impact and of effective emergency planning under both current and future climate conditions in Atlantic Canada. This can be achieved by inclusiveness, with on-site collaboration and secondment of researchers and professionals among organizations. Community resources, which have been increasingly effectively used in environmental and natural scientific investigations in the Canadian Arctic, are still underused in Atlantic Canadian hazard studies. Improved intergovernmental and intragovernmental communication, and more integration of academic and government research efforts, would prove highly beneficial, using the limited financial and personnel resources of four provinces with relatively small populations and research communities.

Academic researchers must become more attuned to the necessity of realistically assessing exposure, ideally in a quantitative but at a minimum in a qualitative fashion. In conjunction, governments and communities should support and recognize efforts to assess exposure as valid contributions to community health, rather than as merely academic exercises.

Atlantic Canada possesses a number of advantages for a successful multi-hazard vulnerability assessment, including relatively small population, vibrant community identity and spirit, and popular recognition and acceptance of government roles. The economic difficulties facing some communities, reluctance to acknowledge certain real difficulties, failure to recognize the significance of natural hazards, and the relatively small scale of some hazard events represent challenges. The progress evident in the past 15 years, however, suggests that it is possible to develop a multi-disciplinary, integrated, multi-hazard approach to sensitivity, exposure, and vulnerability for investigations in Atlantic Canada.

ACKNOWLEDGMENTS

Support for studies referred to in this chapter came from C-CIARN; the Climate Change Action Fund, Natural Resources Canada; Memorial University of Newfoundland; the ministries of Environment and Conservation and of Natural Resources, Newfoundland and Labrador; NSERC; and Public Safety Canada. Our discussions with Martin Batterson, Trevor Bell, Cathy Conrad, Don Forbes, Dave Liverman, Kyle McKenzie, Kathleen Parewick, Kathryn Parlee, John Shaw, Ian Spooner, Bob Taylor, Steve Tomblin, Liette Vasseur, and especially Gail Catto since 1989 have helped greatly as investigations proceeded. Thanks also to the many students whose research has improved our knowledge of natural hazards in Atlantic Canada.

REFERENCES

Adams, G.C. 1993. *Gypsum and Anhydrite in Nova Scotia*. 3rd ed.
Information Circular ME 16. Halifax: Nova Scotia Department of
Natural Resources, Mineral Resources Branch.

Batterson, M., S. McCuaig, and D. Taylor. 2006: "Mapping and assessing
risk of geological hazard on the northeast Avalon Peninsula and
Humber Valley, Newfoundland." *Current Research, Government of
Newfoundland and Labrador, Department of Natural Resources,
Geological Survey, Report 2006–1*: 147–60.

Batterson, M.J., D.G.E. Liverman, J. Ryan, and D. Taylor. 1999. "The assessment of geological hazards and disasters in Newfoundland: an update." *Current Research, Newfoundland Department of Mines and
Energy 99-1*: 95–123.

Belpeiro, A.P. 1993. "Land subsidence and sea level rise in the Port
Adelaide estuary: implications for monitoring the greenhouse effect."
Australian Journal of Earth Sciences 40: 359–68.

Beltaos, S. 1983. "River ice jams: theory, case studies, and applications."
*American Society of Civil Engineering, Journal of Hydraulics Division
109 (HY10)*: 1338–59.

– 1999. "Climatic effects on the changing ice breakup regime of the Saint
John River." In *River Ice Management with a Changing Climate. Pro-
ceedings, 10th Workshop on River Ice, 8–11 June 1999, Winnipeg*.
Sidney, BC: Committee on River Ice Processes and the Environment,
Hydrology Section, Canadian Geophysical Union: 251–64.

Beltaos, S., and B.C. Burrell. 2002. "Extreme ice jam floods along the Saint
John River, New Brunswick." *International Association for Hydrological Science, Publication 271*: 9–14.

Beltaos, S., S. Ismail, and B.C. Burrell. 2003. "Midwinter breakup and
jamming on the upper Saint John River: a case study." *Canadian Journal
of Civil Engineering 30*: 77–88.

Brake, K.K. 2008. "An All-Hazard Assessment of the Marystown Area,
Burin Peninsula, Newfoundland and Labrador." Master of Environmental Science thesis, Memorial University of Newfoundland.

Broster, B.E. 1998. "Aspects of engineering geology at Fredericton, New
Brunswick." In P.F. Karrow and O.L. White, eds., *Urban Geology of
Canadian Cities*. St John's, NL: Geological Association of Canada
Special Paper 42: 401–8.

Bruce, J. 2002. *Consequence Analysis of Storm Surge in the Charlottetown, Prince Edward Island, Area*. Ottawa: Office of Critical Infrastructure and Emergency Preparedness, Government of Canada.

Bryan, W.B., and R.S. Stephens. 1993. "Coastal bench formation at Hanauma Bay, Oahu, Hawaii." *Geological Society of America Bulletin* 105: 377–86.

Burke, K.B.S. 1984: "Earthquake activity in the Maritime provinces." *Geoscience Canada* 11: 16–22.

Canada-Newfoundland Flood Damage Reduction Program. 1990. *Flood Information Map: Rushoon.* St John's, NL: Environment Canada and Newfoundland Department of the Environment.

Catto, N.R. 1998. "Comparative analysis of striations and diamicton fabrics as ice flow direction indicators, Malpeque Bay region, Prince Edward Island." *Boreas* 27: 259–74.

– 2006. "More than 16 years, more than 16 stressors: evolution of a reflective gravel beach 1989–2005." *Géographie physique et Quaternaire* 60: 49–62.

– 2008. *Natural Hazards Mapping, Assessment, and Communication: The Case of Newfoundland and Labrador, Canada.* EGU 2008–A-07300. European Geosciences Union, Vienna, April.

– 2010. *A Review of Academic Literature Related to Climate Change Impacts and Adaptation in Newfoundland and Labrador.* Report to the Cabinet Secretariat Executive, Government of Newfoundland and Labrador.

– and G. Catto. 2009. *Geomorphology and Sedimentology of the Hog Island (Pemamgiag) Sandhills, PEI.* Report to the Mi'kmaq Confederacy of Prince Edward Island.

– E. Edinger, D. Foote, D. Kearney, G. Lines, B. DeYoung, and W. Locke. 2006. *Storm and Wind Impacts on Transportation, SW Newfoundland.* Climate Change Impacts and Adaptations Directorate, Report A 804.

– D. Forbes, and D. Liverman. 2000. "Communicating coastal geoscience to Newfoundland and Labrador communities: failures, successes, and challenges." Abstract. In *GeoCanada 2000.* Calgary: GAC-MAC.

– H. Griffiths, S. Jones, and H. Porter. 2000a. "Late Holocene sea level changes, eastern Newfoundland." In *Current Research, Newfoundland Department of Mines and Energy, Report 2000–1:* 49–59.

– and H. Hickman. 2004. *Flood Sensitivity Analysis for Newfoundland Communities.* Ottawa: Office of Critical Infrastructure Protection and Emergency Preparedness Canada.

– and K. Parewick. 2008. "Hazard and vulnerability assessment and adaptive planning: mutual and multi-lateral community-researcher communication, Arctic and Atlantic Canada." In D. Liverman, ed., *Communicating Geoscience.* Geological Society of London 305: 123–40.

- and L. St Croix. 1998. "Urban geology of St. John's, Newfoundland." In P.F. Karrow and O.L. White, eds., *Urban Geology of Canadian Cities*. St John's, NL: Geoscience Canada: 445–62.
- D. Scruton, and N. Ollerhead. 2003. "The coastline of Eastern Newfoundland." In *Canadian Technical Report of Fisheries and Aquatic Sciences*. Ottawa: Fisheries and Oceans Canada: 2495.
- and S. Tomblin. 2009a. *Response to Natural Hazards: Multi-Level Governance Challenges in Newfoundland and Labrador, Canada*. EGU 2009–5418. IGU Commission for Geosciences Education and European Geosciences Union, Vienna, April.
- 2009b. *Treacherous Terrain, Dark Nature, and Twillicks: Response to Natural Hazards in Newfoundland and Labrador*. Canadian Quaternary Association, Vancouver, May.
- In press. "Hazards, responses, and emergency measures planning: the Newfoundland and Labrador perspective." In R. Young, ed., *Multi-Level Governance in Canada*. Toronto: University of Toronto Press.
Cendrero, A., and D.W. Fischer. 1997. "A procedure for assessing the environmental quality of coastal areas for planning and management." *Journal of Coastal Research* 13: 732–44.
Chang, E.K.M., and Y. Guo. 2007. "Is the number of North Atlantic tropical cyclones significantly underestimated prior to the availability of satellite observations?" *Geophysical Research Letters* 34: L14801.
Connors, S., and C. Tuck. 1999. Integrated Management in Coastal Conception Bay: Preliminary Investigations. Marine Institute of Memorial University. Unpublished report to Department of Fisheries and Oceans, St John's, NL.
Conrad, C.A., and R. Hanson. 2004. "Actual versus perceived risk of hurricanes in Nova Scotia in the year prior to Hurricane Juan and the potential for perception change in a post-Juan period." Abstract, Canadian Association of Geographers, annual meeting, Moncton: 4.
Cooper, J.A.G., and S. McLaughlin. 1998. "Contemporary multidisciplinary approaches to coastal classification and environmental risk analysis." *Journal of Coastal Research* 14: 512–24.
Dawson, A.G. 1999. "Linking tsunami deposits, submarine slides, and offshore earthquakes." *Quaternary International* 60: 119–26.
Edwardson, K.A., D.L. Forbes, J. Shaw, L. Johnston, D. Frobel, and D. Locke. 1993. *Nearshore and Beach Surveys along the Northeast Newfoundland Coast: Dog Bay, Gander Bay, Green Bay, and Baie Verte*. Cruise Report 92–031. Ottawa: Geological Survey of Canada, Open File 2619.

Elsner, J.B., J.P. Kossin, and T.H. Jagger. 2008. "The increasing intensity of the strongest tropical cyclones." *Nature* 455: 92–5.

Emanuel, K., R. Sundararajan, and J. Williams. 2008. "Hurricanes and global warming – results from downscaling IPCC AR4 simulations." *Bulletin, American Meteorological Society* 89: 347–67.

Environment Canada. 1982. *Canadian Climate Normals, 1970–1980.* Ottawa: Atmospheric Environment Service, Environment Canada.

– 1988. *Resource Assessment and Sensitivity Analysis of the South Coast of Newfoundland.* Ottawa: Environment Canada.

– 2004. Hurricane Juan. www.atl.ec.gc.ca/weather/hurricane/juan

Evans, S.G. 1999. *Landslide Disasters in Canada 1840–1998.* Ottawa: Geological Survey of Canada, Open File 3712.

– 2001. "Landslides." In G.R. Brooks, ed., "A synthesis of geological hazards in Canada." *Geological Survey of Canada Bulletin* 548: 43–80.

Finck, P.W. 1993. *An Evaluation of Debris Avalanches in the Central Cape Breton Highlands, Nova Scotia.* Paper 93-1. Halifax: Mines and Energy Branch, Nova Scotia Department of Natural Resources.

Forbes, D., G. Parkes, C. O'Reilly, R. Daigle, R. Taylor, and N. Catto. 2000. "Storm surge, sea-ice, and wave impacts of the 21–22 January 2000 storm in coastal communities of Atlantic Canada." Canadian Meteorological and Oceanographic Society, 34th Congress, Victoria, BC, 29 May–2 June.

Forbes, D.L. 1984. "Coastal geomorphology and sediments of Newfoundland." In *Geological Survey of Canada, Paper 84-1B*: 11–24.

– G.K. Manson, R. Chagnon, S. Solomon, J.J. van der Sanden, and T.L. Lynds. 2002. "Nearshore ice and climate change in the southern Gulf of St. Lawrence." In *Ice in the Environment: Proceedings of the 16th IAHR International Symposium on Ice, Dunedin, New Zealand, 2–6 December, 2002*: 344–51.

– J.D. Orford, R.W.G. Carter, J. Shaw, and S.C. Jennings. 1995. "Morphodynamic evolution, self-organisation, and instability of coarse-clastic barriers on paraglacial coasts." *Marine Geology* 126: 63–85.

– G.S. Parkes, G.K. Manson, and L.A. Ketch. 2004. "Storms and shoreline retreat in the southern Gulf of St. Lawrence." *Marine Geology* 210: 169–204.

Genest, C., and M.-C. Joseph. 1989. "88 centimetres of coastal erosion per year: the case of Kildare (Alberton), Prince Edward Island, Canada." *GeoJournal* 18: 297–303.

Gornitz, V. 1990. "Vulnerability of the east coast, USA, to future sea-level rise." *Journal of Coastal Research* special issue 9: 201–37.

– 1991. "Global coastal hazards from future sea level rise." *Palaeogeogra-phy, Palaeoclimatology, Palaeoecology* 89: 379–98.
– 1993. "Mean sea level changes in the recent past." In R.A. Warrick, E.M. Barrow, and T.M.L. Wigley, eds., *Climate and Sea Level Change: Observations, Projections, and Implications.* Cambridge: Cambridge University Press: 25–44.

Gornitz, V., R.C. Daniels, T.W. White, and K.R. Birdwell. 1993. *The Development of a Coastal Risk Assessment Database: Vulnerability to Sea-Level Rise in the US Southeast.* Washington, DC: US Government Report DE-AC05–84, Environmental Sciences Division Publication 3999.

Gornitz, V., and P. Kanciruk. 1989. "Assessment of global coastal hazards from sea level rise." In *Coastal Zone '89, Proceedings 6th Symposium Coastal and Ocean Management:* 1345–59.

Gornitz, V., T.W. White, and R.M. Cushman. 1991. "Vulnerability of the US to future sea-level rise." In *Coastal Zone '91, Proceedings of the 7th Symposium on Coastal and Ocean Management*, American Society of Civil Engineers: 1345–59.

Grant, D.G. 1994. *Quaternary Geology, Cape Breton Island, Nova Scotia.* Geological Survey of Canada Bulletin 482.

Hare, F.K., R.B.B. Dickison, and S. Ismail. 1997. *Climatic Variation over the Saint John Basin: An Examination of Regional Behaviour.* Climate Change Digest CCD 97–02. Ottawa: Atmospheric Environment Service, Environment Canada.

Hickman, H. 2006. "Flood Hazard and Vulnerability in Newfoundland Communities." MSc thesis, Environmental Science, Memorial University of Newfoundland.

Hilmi, K., T. Murty, M.I. El Sabh, and J.-P. Chanut. 2002. "Long-term and short-term variations of sea level in eastern Canada: a review." *Marine Geodesy* 25: 61–78.

Hirschboeck, K.K., L.L. Ely, and R.A. Maddox. 2000. "Hydroclimatology of meterologic floods." In E.E. Wohl, ed., *Inland Flood Hazards.* Cambridge: Cambridge University Press: 39–72.

House, K, and N.R. Catto. 2008. *Gypsum Karst: Human Initiation and Response, Woodville, NL, Canada.* EGU 2008–A-09878. European Geosciences Union, Vienna, April.

Hyndman, D., D. Hyndman, and N.R. Catto. 2008. *Natural Hazards and Disasters.* Toronto: Nelson.

Ismail, S., and J.L. Davis. 1992. "Ice jams profiling on the Saint John River, New Brunswick." In *Proceedings, 11th International Symposium on Ice, 15–19 June 1992, Banff.* Madrid: International Association for Hydraulic Research: 383–94.

Josenhans, H., and S. Lehman. 1999. "Late glacial stratigraphy and history of the Gulf of St. Lawrence, Canada." *Canadian Journal of Earth
Sciences* 36: 1327–45.

Knight, I. 1983. *Geology of the Carboniferous Bay, St. George Subbasin,
Western Newfoundland.* Government of Newfoundland and Labrador,
Department of Mines and Energy, Mineral Development Division,
Memoir 1: 382.

Kossin, J.P., K.R. Knapp, D.J. Vimont, R.J. Murnane, and B.A. Harper.
2007. "A globally consistent reanalysis of hurricane variability and
trends." *Geophysical Research Letters* 34 (4): 1–6, L04815. doi:10.1029/
2006GL028836

Landsea, C.W., B.A. Harper, K. Hoarau, and J.A. Knaff. 2006. "Can we
detect trends in extreme tropical cyclones?" *Science* 313: 452–4.

Liverman, D., N.R. Catto, and M.J. Batterson. 2006. "Geological hazards
in St. John's." *Newfoundland and Labrador Studies* 21: 71–96.

Liverman, D.G.E., 2008. *Killer Snow.* St John's, NL: Flanker Press.

– M.J. Batterson, and D. Taylor. 2003. "Geological hazards and disasters
in Newfoundland – recent discoveries." Newfoundland Department of
Mines and Energy. In *Geological Survey, Report* 03–1: 273–8.

– M.J. Batterson, D. Taylor, and J. Ryan. 2001. "Geological hazards and
disasters in Newfoundland." *Canadian Geotechnical Journal* 38: 936–56.

– D.L. Forbes, and R.A. Boger. 1994a. "Coastal monitoring on the Avalon
Peninsula." In *Newfoundland Department of Mines and Energy,
Current Research 1994, Geological Survey Branch Report* 94–1: 17–27.

– 1994b. "Coastal monitoring on the Avalon Peninsula, Newfoundland."
In P.G. Wells and P.J. Ricketts, eds., *Coastal Zone Canada 1994, Cooperation in the Coastal Zone, Halifax, Bedford Institute of Oceanography* 5: 2329–44.

Marsh, G.P. 1864. *Man and Nature; or, Physical Geography as Modified
by Human Action.* Reprint 1965. Cambridge, MA: Belknap Press of
Harvard University.

McCulloch, M.M., D.L. Forbes, and R.D. Shaw. 2002. *Coastal Impacts of
Climate Change and Sea-Level Rise on Prince Edward Island.* Ottawa:
Geological Survey of Canada, Open File 4261.

Murdoch, B. 1825. *A Narrative of the Late Fires at Miramichi, New
Brunswick – 1825.* Halifax: P.J. Holland.

Nichols, C. 1995. *Sedimentology, Geomorphology, and Stability of Peter's
River Beach, Avalon Peninsula, Newfoundland.* Contractual report to
Department of Fisheries and Oceans, St John's, NL.

O'Brien, J.P., M.D. Bishop, K.S. Regular, F.A. Bowdring, and T.C.
Anderson. 1998. *Community-Based Coastal Resource Inventories in*

Newfoundland and Labrador: Procedures Manual. St John's: Fisheries and Oceans Canada.

Ollerhead, J., and R.G.D. Davidson-Arnott. 1995. "The evolution of Buctouche Spit, New Brunswick, Canada." *Marine Geology* 124: 215–36.

O'Reilly, C.T., D.L. Forbes, and G.S. Parkes. 2005. "Defining and adapting to coastal hazards in Atlantic Canada: facing the challenge of rising sea levels, storm surges and shoreline erosion in a changing climate." *Ocean Yearbook* 19: 189–207.

Orford, J.D., R.W.G. Carter, and S.C. Jennings. 1996. "Control domains and morphological phases in gravel-dominated coastal barriers of Nova Scotia." *Journal of Coastal Research* 12: 589–604.

– R.W.G. Carter, S.C. Jennings, and A.C. Hinton. 1995. "Processes and timescales by which a coastal gravel-dominated barrier responds geomorphically to sea-level rise: Story Head Barrier, Nova Scotia." *Earth Surface Processes and Landforms* 20: 21–37.

Owens, E.H. 1994. *Coastal Zone Classification System for the National Shoreline Sensitivity Mapping Program.* Ottawa: OCC Limited, Environment Canada.

Paone, L. 2003. "Hazard Sensitivity in Newfoundland Coastal Communities – Impacts and Adaptations to Climate Change: A Case Study of Conception Bay South and Holyrood, Newfoundland." MSc thesis, Department of Geography, Memorial University of Newfoundland.

Pariset, E., R. Hausser, and A. Gagnon. 1966. "Formation of ice covers and ice jams in rivers." *American Society of Civil Engineering, Journal of Hydraulics Division* 92: 1–24.

Park, A.F., and B.E. Broster. 1996. "Influence of glacitectonic fractures on wall failure in open excavations: Heath Steele Mines, New Brunswick, Canada." *Canadian Geotechnical Journal* 33: 720–31.

Parkes, G.S., and L.A. Ketch. 2002. *Storm-Surge Climatology.* Geological Survey of Canada. Open File 4261.

Peddle, P. 2004. "When ice prevails: the Badger, Newfoundland, flood, February 2003." In *Program with Abstracts, 1st Annual Canadian Risks and Hazards Network Symposium, Winnipeg*: 27.

Pennell, C. 1993. "Geomorphology of Biscay Bay Brook, Newfoundland." Honours BSc thesis, Department of Geography, Memorial University of Newfoundland.

Pittman, D. 2004. "Analysis of Coastal Geomorphological Processes on a Boreal Coarse Clastic Barrier: Long Pond Barachois, Conception Bay, Newfoundland." MSc thesis, Department of Geography, Memorial University of Newfoundland.

Quick, M.C., and A. Pipes. 1976. "A combined snowmelt and rainfall run-off model." *Canadian Journal of Civil Engineering* 3: 449–60.

Ruffman, A. 1991. "The 1929 'Grand Banks' earthquake and the historical record of earthquakes and tsunamis in eastern Canada." In *Proceedings, Geological Survey of Canada Workshop on Eastern Seismicity Source Zones for the 1995 Seismic Hazard Maps, March 18–19, 1991.* Ottawa: Geological Survey of Canada. Open File 2437: 371–96.

– 1993. *Reconnaissance Search on the South Coast of the Burin Peninsula, Newfoundland, for Tsunami-Laid Sediments Deposited by the 'Tidal Wave' following the November 18, 1929, Laurentian Slope Earthquake, August 17–September 2, l993.* Geomarine Associates Ltd, Halifax, NS, Project 90–19, Contract Report for Seismology, Lamont-Doherty Earth Observatory of Columbia University, Palisades, NY, Contract No. NRC-04–92–088, US Nuclear Regulatory Commission, Washington, DC, 26 Sept.: 228.

– 1995a. "Comment on: 'The Great Newfoundland Storm of 12 September 1775' by Anne E. Stevens and Michael Staveley." *Bulletin of the Seismological Society of America* 85: 646–9.

– 1995b. *Tsunami Runup Maps as an Emergency Preparedness Planning Tool: The November 18, 1929, Tsunami in St. Lawrence, Newfoundland, As a Case Study.* Geomarine Associates Ltd, Halifax, NS, Project 94–14, for Emergency Preparedness Canada: 399.

– 1996. "The multidisciplinary rediscovery and tracking of 'The Great Newfoundland and Saint-Pierre et Miquelon Hurricane of September 1775.'" *Northern Mariner* 6 (3): 11–23.

Ruitenberg, A.A., and S.R. McCutcheon. 1982. "Acadian and Hercynian structural evolution of southern New Brunswick." In P. St-Julien and J. Beland, eds., *Major Structural Zones and Faults of the Northern Appalachians.* Special Paper 24. St John's: Geological Association of Canada: 131–48.

Ruitenberg, A.A., S.R. McCutcheon, and D.V. Venugopal. 1976. "Recent gravity sliding and coastal erosion, Devils Half Acre, Fundy Park, New Brunswick: geological explanation of an old legend." *Geoscience Canada* 3: 237–9.

Saunders, M.A., and A.S. Lea. 2008. "Large contribution of sea surface warming to recent increase in Atlantic hurricane activity." *Nature* 451: 557–60.

Scott, D.B., and D.A. Greenberg. 1983. "Relative sea level rise and tidal development in the Fundy tidal systems." *Canadian Journal of Earth Sciences* 20: 1554–64.

SGCS (Southern Gulf of St Lawrence Coalition on Sustainability). 2004. *Climate Change and Coastal Communities: Concerns and Challenges for Today and Beyond*. Final Report, workshop, Bouctouche, NB, 11–13 Nov. Southern Gulf of St. Lawrence Coalition on Sustainability.

Shaw, J., and J. Ceman. 1999. "Salt-marsh aggradation in response to late-Holocene sea-level rise at Amherst Point, Nova Scotia, Canada." *Holocene* 9: 439–51.

Shaw, J., P. Gareau, and R.C. Courtney. 2002. "Palaeogeography of Atlantic Canada 13–0 kyr." *Quaternary Science Reviews* 21: 1861–78.

Shaw, J., R.B. Taylor, and D.L. Forbes. 1993. "Impact of the Holocene Transgression on the Atlantic coastline of Nova Scotia." *Géographie physique et Quaternaire* 47: 221–38.

– R.B. Taylor, D.L. Forbes, M.-H. Ruz, and S. Solomon. 1994. "Susceptibility of the Canadian coast to sea-level rise." In P.G. Wells and P.J. Ricketts, eds., *Coastal Zone Canada 1994, Co-operation in the Coastal Zone*. Halifax: Bedford Institute of Oceanography 5: 2377.

– R.B. Taylor, D.L. Forbes, and S. Solomon. 2001. "Sea level rise in Canada." In G.R. Brooks, ed., "A synthesis of geological hazards in Canada." *Geological Survey of Canada Bulletin* 548: 225–6.

– R.B. Taylor, D.L. Forbes, S. Solomon, and M.-H. Ruz. 1998. *Sensitivity of the Coasts of Canada to Sea-Level Rise*. Geological Survey of Canada, Bulletin 505.

Sherin, A.G., and K.A. Edwardson. 1996. "A coastal information system for the Atlantic Provinces of Canada." *Marine Technology Society Journal* 30 (4): 20–7.

Smith, L., D. Liverman, N.R. Catto, and D. Forbes. 2004a. "Coastal hazard vulnerability as a geoindicator, Avalon Peninsula, Newfoundland, Canada." Abstract. Firenze: International Geological Union Congress.

– 2004b. "Coastal hazard vulnerability, Conception Bay South–Holyrood, NL: impacts and adaptations to climate variability." Abstract. St John's: Coastal Zone Canada.

Stevens, A.E. 1983. *Miramichi, New Brunswick, Canada, Earthquake Sequence of 1982: A Preliminary Report*. Ottawa: Earthquake Engineering Research Institute, Energy Mines and Resources.

– 1995. "Reply to comments on 'The Great Newfoundland Storm of 12 September 1775.'" *Bulletin of the Seismological Society of America* 85: 650–2.

– and M. Staveley. 1991. "The great Newfoundland storm of 12 September 1775." *Bulletin of the Seismological Society of America* 81: 1398–402.

Tanoli, S.K., and R.K. Pickerill. 1988. "Lithostratigraphy of the Cambrian–Lower Ordovician Saint John Group, southern New Brunswick." *Canadian Journal of Earth Sciences* 25: 669–90.

Taylor, R.B., D.L. Forbes, D. Frobel, J. Shaw, and G. Parkes. 1996b. "Shoreline response to major storm events in Nova Scotia." In R.W. Shaw, ed., *Climate Change and Climate Variability in Atlantic Canada*. Occasional Paper 9. Dartmouth, NS: Environment Canada, Atlantic Region: 253–67.

– 1997. *Hurricane Hortense Strikes Atlantic Nova Scotia: An Examination of Beach Response and Recovery*. Ottawa: Geological Survey of Canada. Open File 3503.

– H. Josenhans, B.A. Balcom, and A.J.B. Johnston. 2000. *Louisbourg Harbour through Time*. Ottawa: Geological Survey of Canada. Open File 3896.

– J. Shaw, D.L. Forbes, and D. Frobel. 1996a. *Eastern Shore of Nova Scotia: Coastal Response to Sea-Level Rise and Human Interference*. Ottawa: Geological Survey of Canada. Open File 3244.

– S.L. Wittman, M.J. Milne, and S.M. Kober. 1985. *Beach Morphology and Coastal Changes at Selected Sites, Mainland Nova Scotia*. Geological Survey of Canada, Paper 85–12.

Taylor, T. 1994. "Coastal Land Management, Town of Conception Bay South." Honours BA thesis, Department of Geography, Memorial University of Newfoundland.

Thompson, K.R., N.B. Bernier, and P. Chan. 2009. "Extreme sea levels, coastal flooding and climate change with a focus on Atlantic Canada." *Natural Hazards* 51: 139–50.

Vasseur, L., and N. Catto. 2008. "Atlantic Canada." In D.S. Lemmen, F.J. Warren, J. Lacroix, and E. Bush, eds., *From Impacts to Adaptation: Canada in a Changing Climate 2007*. Ottawa: Natural Resources Canada: 119–70.

Watt, W.E. 1989. *Hydrology of Floods in Canada: A Guide to Planning and Design*. Ottawa: Associate Committee on Hydrology, National Research Council of Canada.

Webb, T.C. 2002. *Geology, Development History, and Exploration Alternatives for Gypsum Resources at Cape Maringouin (Parts of NTS 21 H/10, 21 H/15 and 21 H/16), Westmorland County, Southeastern New Brunswick*. Fredericton: Ministry of Natural Resources, New Brunswick. Open File OF-8.

Webster, T.L., D.L. Forbes, E. MacKinnon, and D. Roberts. 2006. "Flood-risk mapping for storm-surge events and sea-level rise using LiDAR for southeast New Brunswick." *Canadian Journal of Remote Sensing* 32: 194–211.

White, J.W. 1902. "Ferryland – what doth not appear in history." *Newfoundland Quarterly* March: 19–21.

White, M. 2002. "A Preliminary Assessment of Slope Stability and Rockfall Hazard, St. Brendan's, Bonavista Bay, Newfoundland." Master of Environmental Science project report, Memorial University of Newfoundland.

Wohl, E.E. 2000. *Inland Flood Hazards*. Cambridge: Cambridge University Press.

Wright, G. 2004. "Coastline Classification and Geomorphic Processes at Ferryland Beach." Honours BSc thesis, Department of Geography, Memorial University of Newfoundland.

11

Infrastructure Failure Interdependencies in Extreme Events: The 1998 Ice Storm

STEPHANIE E. CHANG, TIMOTHY L. McDANIELS, JOEY MIKAWOZ, AND KRISTA PETERSON

INTRODUCTION

The impacts of natural disasters are often greatly prolonged and exacerbated by disruptions to critical infrastructure systems, such as electric power, water, and transportation. These lifelines provide vital services for societal functions. Canadians rely on infrastructure essential to their health, safety, and security, in addition to their economic well-being. The loss of any of it in disasters – whether natural or human-induced – can result in widespread, catastrophic effects and seriously disrupt patterns of human activity (PCCIP 1997; Rinaldi, Peerenboom, and Kelly 2001). In fact, the government of Canada (GOC) defines critical infrastructure as "physical and information technology facilities, networks, services and assets, which if disrupted or destroyed would have a serious impact on the health, safety, security or economic well-being of Canadians or the effective functioning of governments in Canada" (GOC 2004, 5).

Analysts, planners, and decision-makers have begun to recognize that these systems are highly interconnected and mutually interdependent in a variety of ways. As a case in point, the Canadian government argues that "interdependency analysis must be integrated into risk management decisions, mitigation and preparation strategies, and response and recovery" (GOC 2004, 10). Thus the disruption to electric power in a disaster, for instance, is significant not only for its direct impact on society but also in triggering or

exacerbating disruptions to water, transportation, and other systems, which in turn cause further societal impacts.

This chapter sets out a conceptual framework for characterizing
the nature, extent, and severity of the effects of these close and overlapping connections in disasters. Such infrastructure failure interdependencies (IFIs) can be traced back to some initial infrastructure
failure associated with an extreme event. As defined by the (US)
National Science Foundation, extreme events involve non-linear responses, low probabilities, high consequences, and the potential for
systems interaction that leads to catastrophic losses (Stewart and
Bostrom 2001). Our framework emphasizes interdependent failures,
rather than interdependencies (see also McDaniels et al. 2007;
Chang, McDaniels, and Beaubien 2009).

The electric power sector is a particularly important example for
exploring IFIs. The risk of large-scale electrical failures from extreme
events is increasing, because rising demand has not been met by sufficient capacity, leaving these systems more vulnerable to any kind of
system disturbance. In the United States in the 1990s, consumer demand increased 35 per cent, and capacity only 18 per cent. This dis-
crepancy resulted in 41 per cent more large outages in the second
half of the decade (Amin 2004). Infrastructure systems in general are
becoming more congested, making them increasingly vulnerable.
When major power outages affect other infrastructure, these interactions prolong and greatly exacerbate the consequences of the initial
outage. In a real sense, the interaction patterns are the pathways
through which the secondary or indirect effects of a major outage
(because of a natural or human-induced event) ripple through societal interactions and economic activity. We demonstrate this phenomenon by applying the conceptual framework to the ice storm of
1998 in northeastern North America as an IFI in an extreme event.

The first section discusses concepts and a conceptual framework
characterizing IFIs. The second describes our methods for collecting
data on IFIs in disasters. The third applies the framework to the
1998 ice storm.

IFI CONCEPTS AND FRAMEWORK

Concepts of Infrastructure Interdependencies

Rinaldi, Peerenboom, and Kelly (2001) provide a basis for explaining
how interdependent infrastructure systems are, and thus vulnerable

to multiple, sequential failures. They offer dimensions for describing these interdependencies, as well as definitions for these dimensions and related terms. They touch on types of failures within complex infrastructure systems. Finally, they, and others (Ezell, Farr, and Wiese 2000; Thomas et al. 2003), portray infrastructures as complex adaptive systems, with emergent properties that one can discern only by studying the system interactions in aggregate.

This chapter builds directly on the work mentioned above, by assessing specific kinds of IFIs within a defined area of infrastructure systems (typically within a city or region). We explore questions whose answers help make a systemic perspective relevant for risk management decisions:

- What consequences matter most when examining the potential for failures in interconnected infrastructure systems?
- What consequences matter most for decisions about managing these failures?
- How can one define and estimate the likelihood of IFIs in a given context?
- How can one judge the severity of the consequences of IFIs?
- What patterns of IFIs are the most significant sources of concern?

To address these questions, we adopt an empirical approach that seeks to characterize patterns of IFIs and their impact. Rather than beginning with modelling infrastructure systems and their interdependencies, we start by conceptualizing IFIs from the standpoint of their impact on society.

A Framework for Characterizing IFIs

The framework we adopt rests on two principal concepts: objects that characterize the scope and content of the IFI of interest and measures that characterize the severity of a consequence of an IFI. We discuss this framework in terms of infrastructure systems that could be affected, due to interdependencies with other systems, after a large-scale failure in the electrical system. This failure could be the result of an extreme event outside the electrical system, such as the 1998 ice storm. The framework will be applied to this outage in the third section.

The basis for our conceptual framework is the observation that an IFI arising from an outage leads to certain societal consequences.

Thus we classify IFIs according to the characteristics of the outage, the interactions between the power system and the affected system, and the consequences to society. The framework (see Table 11.1) structures our empirical database.

The characteristics of the interactions in Table 11.1 come from concepts in Nojima and Kameda (1996), Peerenboom et al. (2002), and Yao, Xie, and Huo (2004) for understanding infrastructure interdependencies. Peerenboom et al. (2002) elucidate four characteristics – cyber, geographical, logical, and physical – with human decisions playing a particular role in the logical interdependencies.

The IFI typologies of cascading and escalating also derive from these scholars' work. The notions of compound damage propagation and restoration are from the research of Nojima and Kameda (1996) into lifeline interactions in the Kobe earthquake. Yao, Xie, and Huo (2004) use multiple earthquakes to develop their classification of lifeline interactions, containing all the categories used by the other two groups, but with different names. They add substitute interaction – "substitutive" in our framework.

Rinaldi, Peerenboom, and Kelly (2001) distinguish dependence from interdependence – relationships between systems that are unidirectional and bidirectional, respectively. We make no such distinction, except for including feedback that indicates whether a particular IFI has a return effect on the power system. The IFI characteristics of complexity, operational state, and adaptive potential are articulated in Peerenboom et al. (2002).

The final five characteristics in the framework relate to the consequence of the IFI and encompass: duration, number of people affected, severity, spatial extent, and type. These characteristics are crucial for analysing the societal effect of IFIs. The next section contains the preliminary analysis of the IFIs during the 1998 ice storm using these values.

DATA AND METHODS

In order to characterize IFIs in various power outages, we have built a database, applying the characteristics and values in the conceptual framework. This database is populated with data from newspaper articles and the work of other researchers. Each record in the database consists of an observed IFI that was noteworthy enough to report in major media or other reports. This database already contains

Table 11.1
Framework for infrastructure failure interdependencies (IFIs)

Characteristic	Values	Explanation
OUTAGE		
Date	Various	Date outage first occurred
Description of event	Various	General location and description of event
Initiating event	Internal	Mechanical or human error within power system
	External	Event outside power system such as a severe storm
Spatial extent	Local	One city or area affected
	Regional	More than one city or area within a province or state affected
	National	More than one state or province affected
	International	More than one country affected
Duration	Minutes, hours, days, weeks	Approximate time it took to restore power to 90 per cent of affected customers
Weather conditions	Moderate	Normal weather conditions for season
	Extreme	Unusual weather conditions such as severe storm
Temperature	Cold	Below 32°F/0°C
	Hot	Above 90°F/32°C
	Mild	Between 32°F and 90°F/0°C and 32°C
INTERDEPENDENCIES		
Affected system	Building support, business, education, emergency services, finance, food supply, government, health care, telecommunications, transportation, utilities	System affected by power outage
Specific system	Various	Subdivisions of affected systems
Description	Various	Brief summary of impact on system
Type of interdependence	Physical	System requires electricity to operate.
	Geographical	System co-located with electrical infrastructure
	Cyber	System linked to electrical system electronically or through information sharing
	Logical	System dependent on electrical system in way that is not physical, cyber, or geographical

Table 11.1 (*Continued*)

Characteristic	Values	Explanation
Type of IFI	Cascading	Disruption of power system directly disrupts affected system.
	Escalating	Disruption of power system exacerbates existing disruption in affected system, increasing severity or outage time.
	Restoration	Power outage hampers restoration of affected system.
	Compound damage propagation	Power system disruption causes serious damage/accidents/problems in affected system.
	Substitutive	System disrupted by demands to substitute for power system
Order	Direct	IFI a direct result of power outage
	Second-order	Power outage once removed as cause of system disruption
	Higher-order	Power outage twice or more removed as cause of system disruption
System failure leading to this effect	See Affected system for list	Electrical in direct-order events; system that caused disruption in affected system in second- and higher-order events
Complexity	Linear	Expected and familiar interactions
	Complex	Unplanned or unexpected sequences of events
Feedback	Yes	Affected system impacts power system.
	No	Affected system doesn't impact power system.
Operational state	At capacity	Affected system operating at 100 per cent when power outage occurred
	Near capacity	Affected system operating above 90 per cent when power outage occurred
	Below capacity	Affected system operating at 90 per cent or below when power outage occurred
Adaptive potential	High	System that can respond quickly in a crisis
	Low	Inflexible system that can't respond quickly
Restart time	Minutes, hours, days, weeks	Time for affected system to return to pre-outage operating capacity once electric power restored
CONSEQUENCE		
Severity	Minor	Minor modifications in daily routine or plans that cause negligible hardship to person or entity
	Moderate	A few modifications in daily routine or plans that cause some hardship to person or entity
	Major	Significant modifications in daily routine or plans that cause considerable hardship to person or entity

Table 11.1 (*Continued*)

Characteristic	Values	Explanation
Type	Economic, health, safety, social, environmental	Primary category under which consequence falls
Spatial extent	Local	One city or area affected
	Regional	More than one city or area within a province or state affected
	National	More than one state or province affected
	International	More than one country affected
Number of people	Few	In spatial extent of consequence, one neighbourhood or isolated individuals affected
	Many	In spatial extent of consequence, up to half of population affected
	Most	In spatial extent of consequence, at least half of population affected
Duration	Minutes, hours, days, weeks	Duration of consequence, which may be greater than restart time

hundreds of IFIs from three major events – the 1995 earthquake in Kobe, Japan; the 1998 ice storm in northeastern North America; and the blackout of August 2003 also in the northeast. Human and mechanical failures caused the 2003 blackout, unlike the other events, where natural hazards initiated the outages. The major data sources for the ice storm – the focus of our empirical analysis – are the *Montreal Gazette* and the *Ottawa Citizen*.

In order to analyse the data, we developed indices of consequences. The database includes five dimensions of IFI consequences: duration, number of people affected, severity, spatial extent, and type. Except for type of consequence, we weighted consequence levels (Table 11.2).

Using these weights, we assigned each IFI an impact value and an extent value. The impact value (ranging from 1 to 9) is the product of the IFI's duration and severity; high values indicate severe consequences with long duration. The extent value (ranging from 1 to 9) is the product of the IFI's spatial extent and affected population; high values indicate many people affected over a large area.

ANALYSIS AND DISCUSSION

Our preliminary analysis focuses on the 1998 ice storm, which left 4.7 million people in Canada, or 16 per cent of the population, without power for hours, days, or weeks at the worst of the storm

Table 11.2
Weights for consequence levels

	Consequence characteristic		
	Weight = 1	Weight = 2	Weight = 3
Duration	Hours	Days	Weeks
Severity	Minor	Moderate	Major
Spatial extent	Local	Regional	International
No. of people affected	Few	Many	Most

(Lecomte, Pang, and Russell 1998). Purcell and Fyfe (1998) note the storm demonstrated how dependent Canadian society is on electrical power and how vulnerable it is to power outages.

The empirical analysis aims to identify which sectors and IFIs affected society most. To this end, we identified IFIs that caused the most severe disruption, using the consequence indices we described above.

Impact of the IFI

Developing strategies for mitigation requires an in-depth look at the effects of specific IFIs. Each sector includes many sub-sectors, and each of these can potentially experience different types of IFIs. For example, transportation includes sub-sectors such as mass transit and road transportation. IFIs for road transportation may include non-functioning traffic signals and non-operational gas stations. In order to identify the most significant IFIs from a societal point of view, we needed succinct indices that synthesize the multi-dimensional data on impacts. These indices can then help us classify IFIs by significance.

Figure 11.1 plots IFIs in the 1998 ice storm according to their impact and extent indices (defined above). There are 106 IFIs in our database from this event. Each represents a mention of some form of disruption to a dependent infrastructure sector and the associated societal impact. IFIs occurred in 11 sectors: building support, business, education, emergency services, finance, food supply, government, health care, telecommunications, transportation, and utilities. These are almost identical to the Canadian government's list of critical infrastructure sectors: communications and information technology, energy and utilities, finance, food, government, health care, manufacturing safety, transportation, and water. Each IFI record is

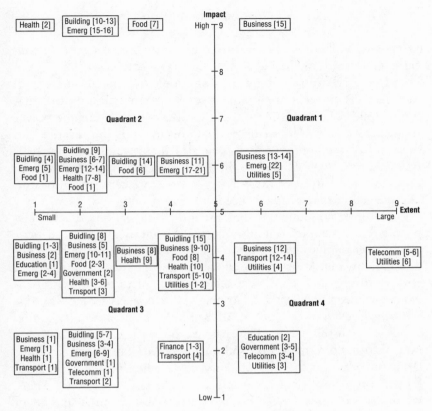

Figure 11.1
Impact and extent of electric-power infrastructure failure interdependencies (IFIs),
1998 ice storm

labelled by a sector and identifier number; for example, "Building
Support [13]" refers to a data point characterized as the lack of electrical heating that led to four deaths from hypothermia. Figure 11.1
plots IFIs with the same index values as a single point with multiple
record labels. Table 11.3 lists recorded IFIs in the database and their
index values.

Figure 11.1 also separates IFIs into four quadrants or categories.
Axes separating the quadrants are at the respective midpoint values
of the potential range of Impact and Extent index values (i.e., 5 on a
scale of 1–9). Quadrant 1 represents major disturbances to a majority of the population; quadrant 2, major disturbances to a small percentage; quadrant 3, minor inconveniences to a small percentage;
and quadrant 4, minor inconveniences to a large percentage.

IFIs of Greatest Societal Concern

From a societal point of view, IFIs in quadrant 1 are of greatest concern. This quadrant includes IFIs that have both high impact and broad extent of impact. In the ice storm, only one IFI (Business [15]) met these criteria: major employers south of Montreal reported that they would shut down for up to two weeks because of the power outage. Not only were many people affected, but the severity and duration of the consequences were also high. Manufacturing was one sector that contributed to the short-term loss of $1.6 million to the economic output of the country – a 0.2 per cent loss in overall real gross domestic product (Lecomte et al. 1998).

Quadrant 2 contains IFIs that had a high impact but affected few people. For example, several Montreal hospitals experienced periodic power outages lasting up to several hours, and many had contaminated water. Major hospital routines were also disrupted (e.g., elective surgery and clinics cancelled; ambulances too busy to provide inter-hospital transfers). Farmers suffered an estimated $14 million in lost revenues in Quebec and $11 million in Ontario (Lecomte, Pang, and Russell 1998). The reasons for these agricultural losses ranged from no refrigeration and no heating and improper air circulation in barns to the loss of dairy cows when milking machines failed (some cows died because they could not be milked). There were also countless other problems stemming from heat system failures. Four people died from hypothermia, and seven from carbon monoxide poisoning while using poorly ventilated heating. Emergency crews (especially firefighters) also became strained when many houses and apartments caught fire; five people died during these fires. A total of 28 deaths was attributed to the ice storm (Lecomte, Pang, and Russell 1998). As we see in Figure 11.1, this quadrant contains more records than any other.

Quadrant 3 contains IFIs that caused inconveniences to a small portion of the population. Examples included traffic light failures (in addition to poor weather) that decreased road safety and hampered repairs. Furthermore, road conditions were unknown due to the communication failures. Many people experienced basement flooding from melting ice, and numerous households that suffered bursting pipes were unable to react when sump pumps could not be powered up. Burst pipes were the third most expensive item for insurers. Consumers were affected by businesses that ran low on high-demand items (e.g., batteries, candles, emergency supplies, and fireplace logs) and by businesses that engaged in price gouging. Quadrant 4 contains IFIs that inconvenienced many people but with an impact not as severe and/or long-lasting as those in quadrant 1. More notable inconveniences included massive school closures and fuel shortages when two major oil refineries closed temporarily. During this time, consumers were also unable to pump remaining fuel. Power outages caused Hydro Québec to request voluntary closures from business in order to accelerate repairs. Water pressure became a concern because of fires from improper heating. As water plants had only four to six hours of clean water left, the general public was issued a water-boiling advisory. Perhaps the most time-consuming problem was associated with telecommunication challenges. Public fears were exacerbated through intermittent or poor communication.

CONCLUSIONS

Electric power systems constitute a fundamental infrastructure in modern society (Amin 2004). Virtually every crucial economic and social function depends on the secure, reliable operation of infrastructures. The ice storm of 1998 demonstrated the wide array of system and interdependent infrastructure failures that can stem from natural disasters. Increasing societal resilience to disasters requires not only understanding and mitigating the risks to individual infrastructure systems, but also mitigating how failures in one system can lead to disruptions in others, hamper response and recovery, or result from common-cause failures. Understanding such interdependencies is necessary if planners, operators, and emergency response personnel are to deal effectively with infrastructure disruptions (Peerenboom et al. 2002).

This chapter presents the conceptual basis, approach, and initial progress of an investigation into infrastructure failure interdependencies (IFIs) caused by electric power outages. Strategic approaches are needed to guide community preparedness for future major outages and their disruptive effects. This study seeks to identify the most important IFIs and promising mitigation strategies for them. It makes three advances towards these goals. First, it presents a conceptual framework for characterizing IFIs on the basis of societal, rather than technical, considerations. Second, it develops a unique database of IFIs observed in a major Canadian disaster, the 1998 ice storm. Third, it demonstrates how the framework can be applied to identify IFIs of greatest societal concern. These represent potential focal points for pre-disaster mitigation and preparedness.

The preliminary analysis presented here suggests several areas for further research. The analysis does not incorporate weights or value judgments across types of IFI impacts. From a policy perspective, severe economic disruptions such as temporary business closures may not be equivalent, for example, to severe safety impacts such as deaths. Frameworks for addressing different types of consequences are needed. Also, our empirical approach can complement probabilistic, systems-based, and simulation models of power outages and their effects. Such models may identify low-probability, high-consequence events and impacts that could potentially occur but have not yet done so and therefore would not be amenable to empirical analysis. Another research need is a robust empirical basis that incorporates experiences across a range of event and community types. Commonalities and differences in IFI occurrence across types of natural, technological, and wilful disasters should be explored; for example, IFIs that occur in many types of events should be identified, as they would be promising targets of mitigation from a multi-hazard perspective. Further, while this study focuses on IFIs deriving from electric power failure, the framework can be readily extended to assess other types of infrastructure interdependencies and to set priorities about potential ways to mitigate the likelihood and consequences of their interdependent failures.

REFERENCES

Amin, M. 2004. "North American electricity infrastructure: system security, quality, reliability, availability, and efficiency challenges and their societal impacts." In *National Science Foundation Report on Continuing Crises in National Transmission Infrastructure: Impacts and Options for Modernization.* Minneapolis, MN: chap. 2, 1–20.

Chang, S.E., T. McDaniels, and C. Beaubien. 2009. "Societal impacts of infrastructure failure interdependencies: building an empirical knowledge base." *Proceedings of the 2009 Technical Council on Lifeline Earthquake Engineering (TCLEE) Conference.* Oakland, CA: 693–702.

Ezell, B.C., J.V. Farr, and I. Wiese. 2000. "Infrastructure risk analysis of municipal water distribution system." *Journal of Infrastructure Systems* 6 (3): 118–22.

GOC (Government of Canada). 2004. *Position Paper on a National Strategy for Critical Infrastructure Protection.* www.ocipep.gc.ca/critical/nciap/NSCIP_e.pdf

Lecomte, E.L., A.W. Pang, and J.W. Russell. 1998. *Ice Storm '98.* Institute for Catastrophic Loss Reduction, Research Paper Series No. 1. Toronto.

McDaniels, T., S. Chang, K. Peterson, J. Mikawoz, and D. Reed. 2007. "Empirical framework for characterizing infrastructure failure interdependencies." *Journal of Infrastructure Systems* 13 (3): 175–84.

Nojima, N., and H. Kameda. 1996. "Lifeline interactions in the Hanshin-Awaji earthquake disaster." In *Committee of Earthquake Engineering, the 1995 Hyogoken-Nanbu Earthquake: Investigation into Damage to Civil Engineering Structures.* Tokyo: Society of Civil Engineers: 253–64.

PCCIP (President's Commission on Critical Infrastructure Protection). 1997. *Critical Foundations: Protecting America's Infrastructure.* Washington, DC.

Peerenboom, J., R.E. Fisher, S.M. Rinaldi, and T.K. Kelly. 2002. "Studying the chain reaction." *Electric Perspectives* Jan.–Feb.: 22–35.

Purcell, M., and S. Fyfe. 1998. *Queen's University Ice Storm '98 Study: Emergency Preparedness and Response Issues.* Ottawa: Emergency Preparedness Canada.

Rinaldi, S.M., J.P. Peerenboom, and T.K. Kelly. 2001. "Critical infrastructure interdependencies." *IEEE Control Systems Magazine* Dec.: 11–25.

Stewart, T.R., and A. Bostrom. 2001. *Extreme Event Decision Making Workshop Report.* Arlington, VA. www.albany.edu/cpr/xedm/

Thomas, W.H., M.J. North, C.M. Macal, and J.P. Peerenboom. 2003. "From physics to finances: complex adaptive systems representation of infrastructure interdependencies." In *Dahlgren Division Technical Digest.* Washington, DC: Naval Surface Warfare Center: 58–67.

Yao, B., L. Xie, and E. Huo. 2004. "Study effect of lifeline interaction under seismic conditions." In *Proceedings of the 13th World Conference on Earthquake Engineering.* Vancouver.

Table 11.3
Recorded IFIs in the database and their index values

No.	Specific system	Description	Extent index	Impact index
BUILDING SUPPORT				
1	HVAC	Thirty-seven elderly women evacuated from seniors' residence in downtown Montreal	1	4
2	HVAC	Fifty-five elderly or infirm people without power or heat refuse to leave homes.	1	4
3	Security	Thieves hit several homes knowing alarm system down.	1	4
4	HVAC	Emergency generator provides light, but no heat, for 70 residents, many bedridden.	1	6
5	Garage door	People unable to use vehicles because unable to open electric garage doors	2	2
6	HVAC	Provisions for looking after pets inadequate	2	2
7	Plumbing	Plumbers extremely busy responding to calls from people who want pipes drained	2	2
8	Plumbing	If power returns and water heater tank empty, it could burn out internal heater in tank.	2	4
9	HVAC	More than 100 reported cases of carbon monoxide poisoning	2	6
10	HVAC	Some houses continually burning wood in fireplaces catch fire.	2	9
11	HVAC	Four people die of hypothermia.	2	9
12	HVAC	Several people die in fires from makeshift heating.	2	9
13	HVAC	Six people die of carbon monoxide poisoning from fumes from gas and oil heaters.	2	9
14	HVAC	Carbon monoxide build-up leaves more than 200 residents homeless.	3	6
15	Plumbing	People experience flooding after sump pumps rendered inoperable by power outages.	4	4
BUSINESS				
1	Other	Power outage at Mount Royal Crematorium means bodies not incinerated for several days	1	2
2	Retail	Two businesses looted overnight.	1	4
3	Retail	Direct debit not always operational, some stores accepting IOUs	2	2
4	Retail	Line-ups of an hour for firewood and kerosene	2	2
5	Restaurants	Restaurants powered by generators begin to run out of food.	2	4
6	All	People laid off as businesses close.	2	6

Table 11.3 (*Continued*)

No.	Specific system	Description	Extent index	Impact index
7	Retail	Most stores have exhausted supply of bottled water.	2	6
8	Insurance	Burst pipes expected to be third most expensive item for insurers.	3	4
9	Insurance	Second-highest cost for insurers: replace food gone bad.	4	4
10	Retail	Crowded stores of people in search of batteries, candles, emergency supplies, and fireplace logs.	4	4
11	Hotels	Hotels and motels booked solid by people looking for shelter.	4	6
12	Retail	Many businesses charge extra on high-demand items.	6	4
13	All	Hydro-Québec officials call on shops, businesses, and industry to shut down.	6	6
14	Retail	In short supply: batteries, flashlights, candles, propane, drinking water, etc.	6	6
15	All	Major employers south of Montreal shut down for up to two weeks.	6	9

EDUCATION

No.	Specific system	Description	Extent index	Impact index
1	Daycare	Daycare with no heat and lack of food	1 .	4
2	Schools	Virtually every school board in and around Montreal cancels classes for up to week.	6	2

EMERGENCY SERVICES

No.	Specific system	Description	Extent index	Impact index
1	Shelters	Person moved from shelter to hospital after suffering anxiety attack.	1	2
2	Police	Police report apparent surge in domestic violence.	1	4
3	Police	Due to vandalism and break-in, Hudson imposes 8 p.m. curfew on youth.	1	4
4	Shelters	Frayed nerves and overcrowding affect more than 2,000 ice storm refugees.	1	4
5	Police	Local police station without power	1	6
6	9-1-1	Exceptionally high call volume, partly due to weather.	2	2
7	All	RDI (Radio Canada) news reports that all of Montreal will be blacked out create panic as people leave and emergency lines are flooded.	2	2
8	Police	Three times as many police on duty, many of them patrolling streets (because of looting)	2	2
9	Police	Police in Montreal begin exercising powers to force people at health risk to leave frigid homes.	2	2

Table 11.3 (Continued)

No.	Specific system	Description	Extent index	Impact index
10	Fire	Firefighters busy pumping water out of basements where sump pumps stopped	2	4
11	Shelters	Shelters at Claude Robillard sports centre and Little Burgundy sports complex close after losing power.	2	4
12	All	Without computers, impossible to obtain vital information, including phone numbers and addresses of emergency responders and vulnerable people, civic address lists, maps, and prescription information	2	6
13	All	Many people refuse to leave homes, raising concerns over food, water, and medication safety.	2	6
14	Fire	Several fires caused by improper heating and power surges leave many people homeless.	2	6
15	Police/fire	Three "suspicious fires" reported at rooming-house, restaurant, and old factory	2	9
16	Fire	Several fires caused by improper heating and power surges take lives.	2	9
17	All	Some backup power systems not working because of improper maintenance	4	6
18	Fire	As a result of fire, water pressure a concern for firefighters	4	6
19	Shelters	Many of 200 shelters across Quebec running out of supplies	4	6
20	Shelters	Six of Montreal's shelters filled to capacity	4	6
21	Shelters	Tens of thousands of people forced into emergency shelters	4	6
22	All	Communication was most time-consuming problem	6	6
FINANCE				
1	ATM	Several bank machines out of service, no backup power	4	2
2	Banks	Many banks and credit unions close because of lack of power.	4	2
3	Credit cards	Credit cards not accepted everywhere	4	2
FOOD SUPPLY				
1	Production	Farmers working furiously to protect livestock	1	6
2	Production	Automatic feeders on dairy farms dormant	2	4
3	Storage	92,000 bushels of apples to rot if generator cannot circulate air	2	4

Table 11.3 (*Continued*)

No.	Specific system	Description	Extent index	Impact index
4	Storage	Thousands of chickens die from suffocation.	2	6
5	Production	Getting water for dairy herd a challenge with electric pumps not working	2	6
6	Storage	13.5 million litres of milk dumped	3	6
7	Production	Cows can suffer severe infections and supply dry-up if they're not milked.	3	9
8	Storage	Food stores without backup generators watch perishables wilt, rot, or sour.	4	4

GOVERNMENT

No.	Specific system	Description	Extent index	Impact index
1	Services	Most museums and art galleries and many libraries in city close indefinitely.	2	2
2	Services	Montreal's municipal court closes, but employees are expected to show up for work.	2	4
3	Offices	Most governmental offices closed	6	2
4	Services	Canada Post employees not to report to work, to help restore electricity by closing of high-tech sorting stations	6	2
5	Services	Canada Post offices on Montreal Island and South Shore close.	6	2

HEALTH CARE

No.	Specific system	Description	Extent index	Impact index
1	Hospitals	Medication distributed by staff hiking stairs between hospital's six floors	1	2
2	Public health	In residence for handicapped adults, fireplace overheats and starts fire, which kills one resident and sends three to hospital.	1	9
3	Hospitals	Patients staying to avoid blacked-out homes tying up beds	2	4
4	Hospitals	Crowding leads to influenza spread, affecting hospitals.	2	4
5	Public health	American Red Cross issues urgent call for blood donations.	2	4
6	Public health	Crowding in shelters leads to influenza spread.	2	4
7	Hospitals	Jewish General Hospital (JGH) calls for volunteer nurses; Montreal GH and Royal Victoria H look for volunteers with no experience.	2	6
8	Public health	Elderly tenants (apartment buildings) without power for oxygen tanks and other medical equipment	2	6
9	Public health	Many pharmacies close, unable to refill medication.	3	4
10	Hospitals	Five health care centres rely on emergency power.	4	4

Table 11.3 (*Continued*)

No.	Specific system	Description	Extent index	Impact index
TELECOMMUNICATIONS				
1	Cable	Videotron service interrupted, but batteries last only three to five hours.	2	2
2	Land line telephones	Theft of Bell generator in Lachute area deprives residents of phone service for week.	2	6
3	Media	Radio station dead; not enough fuel to power generators	6	2
4	Media	Daily newspapers unable to publish the day before	6	2
5	All	Communication most time-consuming problem, weakest link	9	4
6	Land line telephones	Telephone service to Bell Canada customers knocked out	9	4
TRANSPORTATION				
1	Roads	Motorist with flat tire unable to locate station open	1	2
2	Taxi	Taxi-drivers scrambling to find gas stations to work through night	2	2
3	Roads	Traffic lights off, causing dozens of fender-benders	2	4
4	Mass transit	Closure of some Métro stations causes long line-ups at bus stops.	4	2
5	Air	Dorval airport loses principal power feed, is low on jet fuel.	4	4
6	Air	Flights cancelled or delayed because of line-ups of planes awaiting de-icing	4	4
7	Mass transit	Montreal South Shore Transit cancels 46 rush-hour bus routes due to power outage.	4	4
8	Roads	Consumers drive for miles to find functioning gas station.	4	4
9	Roads	Road conditions unknown without communication	4	4
10	Mass transit	Limited bus service	4	4
11	Gas stations	Most gas stations unable to pump fuel	6	4
12	Air	Major airlines cancel 255 flights.	6	4
13	Rail	Rail lines shut down: crossing signals, operating signals, and switches no longer working.	6	4
14	Roads	Traffic lights out in blacked-out zones, and reduced visibility	6	4

Table 11.3 (*Continued*)

No.	Specific system	Description	Extent index	Impact index
UTILITIES				
1	Electricity	Ontario Hydro to pay $150 million in repairs due to ice storm	4	4
2	Electricity	Hydro Quebec to pay $600 million in repairs due to ice storm	4	4
3	Water	Atwater and Desbaillets water plants: four to six hours of clean water in reservoirs	6	2
4	Water	As result of fire, water pressure a concern	6	4
5	Oil	Fuel critical because of no pumping power	6	6
6	Oil	Montreal's two east-end oil refineries and two fuel terminals knocked out	9	4

Contributors

LIANNE M. BELLISARIO is a senior policy advisor with the Public Health Agency of Canada (PHAC), Ottawa, Ontario. Her research and policy-related contributions have been in climate change impact and adaptation, the environment, natural resource management, public health, and public safety. E-mail address: lianne.bellisario@ phac-aspc.gc.ca

FIKRET BERKES is professor at the Natural Resources Institute at the University of Manitoba, Winnipeg. He holds the Canada Research Chair in Community-Based Resource Management and the title of distinguished professor. He has published extensively, including *Linking Social and Ecological Systems* (1998), *Sacred Ecology* (1999), and *Breaking Ice* (2005). His work combines social and natural sciences and has been mainly in commons theory and the interrelations between societies and their resources. E-mail address: berkes@cc.umanitoba.ca

MIHIR R. BHATT is director of the All India Disaster Mitigation Institute (AIDMI) in Ahmedabad. He studied and practiced architecture and city planning in India and the United States (at the Massachusetts Institute of Technology). He received the Russell E. Train Institutional Fellowship from the World Wildlife Fund (WWF) in 1997 and the Eisenhower Fellowship in 2000. Currently he is leading implementation of community-based programs to reduce risk and disaster and improve the lives of the poor in India and elsewhere. E-mail address: mihir@aidmi.org

NORM R. CATTO is professor of geography at Memorial University of Newfoundland, St John's. His work focuses on coastal geomorphology, flood risk assessment, Quaternary chronology, and sedimentology. His current interests include coastal management in Newfoundland, community perception of and response to climate change, effects of climate change on transportation, flood sensitivity of communities, and natural hazards in Arctic Canada. E-mail address: ncatto@mun.ca

STEPHANIE E. CHANG is professor with the School of Community and Regional Planning at the University of British Columbia in Vancouver, where she is Canada Research Chair in Disaster Management and Urban Sustainability. Much of her work aims to bridge the gap between engineering, the natural sciences, and the social sciences in addressing the complex issues of natural disasters. Her current research examines communities' disaster resilience and sustainability, mitigation of risks to infrastructure systems, and urban disaster recovery. She won the 2001 Shah Family Innovation prize from the Earthquake Engineering Research Institute and serves on the editorial board of *Earthquake Spectra*. E-mail address: sechang@interchange.ubc.ca

SAYEDUR R. CHOWDHURY received his BSC and MSC in marine science from the University of Chittagong, Bangladesh, and his MSC in GIS and remote sensing from the Asian Institute of Technology, Thailand. He teaches at the Institute of Marine Sciences and Fisheries at the University of Chittagong and is currently pursuing his PhD at the Natural Resources Institute, University of Manitoba, Winnipeg. E-mail address: sayedurrchy@yahoo.com

JULIE DEKENS-KARAMI is project officer at the International Institute for Sustainable Development in Geneva. Her research and policy-related contributions have been in adaptation to climate change, local knowledge about disaster risk, and reducing disaster risk. E-mail address: jkarami@iisd.org

PARNALI DHAR CHOWDHURY is pursuing her PhD in natural resources and environmental management at the University of Manitoba, Winnipeg. She has earned her master of environment degree there and her master of environmental science from

Visva-Bharati University in India. She has published on climate change and human health hazards, ecosystem health, and risk perception. Currently she is working on ecosystem approaches to human health, with a focus on vector-borne diseases. E-mail address: umdharch@cc.umanitoba.ca

DAVID ETKIN is coordinator of the Program in Emergency Management and assistant professor of emergency management at York University, Toronto. During the 1990s, he worked with the Adaptation and Impacts Research Group of the Meteorological Service of Canada, specializing in the interdisciplinary study of natural hazards and disasters, and with the Institute of Environment Studies at the University of Toronto to conduct research on natural hazards and disasters. He led the first Canadian National Assessment of Natural Hazards project and served as co-president of the Canadian Risk and Hazards Network (CRHNet) during 1993–2007. He has published extensively on disasters and natural hazards. E-mail address: etkin@yorku.ca

JAMES S. GARDNER is professor emeritus in the Natural Resources Institute, University of Manitoba, Winnipeg. He has served as vice-president (academic) and provost at Manitoba and as dean of graduate studies at the University of Waterloo, Waterloo, Ontario. Throughout his academic career at Manitoba, Waterloo, and Iowa, he has taught and undertaken research in geomorphology, glaciology, hazards and resource management, and hydrology in mountain regions, with regional foci on the Canadian Cordillera and the greater Himalayas. E-mail address: gardner@cc.umanitoba.ca

POLAT GÜLKAN is professor of civil engineering at Middle East Technical University, Ankara, Turkey. He is an earthquake structural engineer and has been active in many areas of global hazard mitigation since 1971. He served 1996–2004 on the board of directors of the International Association for Earthquake Engineering. He has published extensively, especially on policy and planning aspects of earthquake hazard and on structural engineering. E-mail address: a03516@metu.edu.tr

C. EMDAD HAQUE is director and professor at the Natural Resources Institute at the University of Manitoba, Winnipeg. He has published extensively, including *Hazards in a Fickle Environment* (1997) and *Mitigation of Natural Hazards and Disasters* (2005). He has worked on disaster management, energy policy and sustainability, hazard prevention and mitigation, human responses to floods, the impact of and adaptation to climate change and its associated risk, perception of drought and heat wave risk, and water resource infrastructure and management. He was founding president (2003–8) of the Canadian Risk and Hazards Network (CRHNet) and is a member of the editorial board of *Natural Hazards*. E-mail address: haquece@cc.umanitoba.ca

A. NURAY KARANCI is a lecturer in the Department of Psychology at Middle East Technical University (METU), Ankara, Turkey. In England, she earned her MS in clinical psychology from Liverpool University and her PhD from Hull University. Since 1993, she has been active in research on community participation, disaster management, and psychological consequences of earthquakes. She is the director of Disaster Research and Implementation Centre at METU. E-mail address: karanci@metu.edu.tr

MIZAN R. KHAN is professor of environmental science and management at North South University, Dhaka, Bangladesh. He earned his PhD in environmental policy and management from the School of Public Policy, University of Maryland at College Park. His current research and publications relate to climate change policy, crop insurance, environmental governance, and resource and environmental management. He served as vice-chair of the Least Developed Country (LDC) expert group under the *United Nations Framework Convention on Climate Change* (UNFCCC). In recent years, he has represented Bangladesh in negotiations on global climate change. E-mail address: mizanrk@northsouth.edu

GORDON A. MCBEAN is professor of political science and geography and chair of policy at the Institute of Catastrophic Loss Reduction at the University of Western Ontario in London. He taught 1988–94 at the University of British Columbia in Vancouver and served 1994–2000 as assistant deputy minister for the Meteorological Service of Environment Canada. He has received the Patterson Medal for distinguished contributions to meteorology and is a fellow of the Royal Society of Canada. His research interests are atmospheric and climate sciences, ranging from the natural sciences to the policies of governments and people's responses to them. E-mail address: gmcbean@uwo.ca

TIMOTHY L. McDANIELS is a professor of the University of British Columbia, Vancouver. He has published extensively. His research combines decision sciences and policy analysis, particularly in managing environmental and technology-related societal risks, and climate change adaptation in linked human/ecological systems. E-mail address: timmcd@interchange.ubc.ca

JACK McGEE has since 2000 been president of the Justice Institute of British Columbia. He was dean of science and technology at George Brown College in Toronto and president of St Clair College, Windsor, Ontario. He received an honorary doctorate of laws from the University of Windsor, Ontario. He has served on a variety of business, civic, and educational bodies representing charities and community, provincial, national, and international interests. E-mail address: jmcgee@jibc.bc.ca

JOEY MIKAWOZ is a graduate research assistant of the School of Community and Regional Planning at the University of British Columbia, Vancouver.

BRENDA L. MURPHY is an associate professor of contemporary studies at the Brantford, Ontario, campus of Wilfrid Laurier University. Her current teaching and research interests include emergency management, risk (climate change, disasters, hazardous facility siting), social capital, and environmental justice issues that affect communities and the world. Her case studies included the *E. coli* disaster in Walkerton, Ontario; management of nuclear fuel waste in Canada, Sweden, and the United States; the Pine Lake tornado in Alberta; and the 2003 blackout on the eastern seaboard of North America. E-mail address: bmurphy@wlu.ca

NIRU NIRUPAMA is an associate professor of disaster and emergency management at the School of Administrative Studies, York University, Toronto. She studied statistics at the Indian Institute of Technology, Kanpur, and hydrology at the Indian Institute of Technology, Roorkee, and earned the DENg in water resource engineering from Kyoto University in Japan. Her current teaching and research interests include business continuity, disaster management using GIS and remote sensing, disaster risk management, multi-criteria decision-making using fuzzy logic, and physical dynamics of natural hazards. E-mail address: nirupama@yorku.ca

KRISTA PETERSON is a graduate research assistant of the School of Community and Regional Planning at the University of British Columbia, Vancouver.

TOMMY REYNOLDS was a program associate with the All India Disaster Mitigation Institute (AIDMI) in Ahmedabad. He obtained his master's degree in public administration from Rutgers University in New Jersey. His work has combined project planning and evaluation. E-mail address: tommyjoereynolds@gmail.com.

GRAHAM SMITH has earned his master's degree in natural resource management from the Natural Resource Institute, University of Manitoba, Winnipeg. He is currently investigating gaps between public and expert perceptions of the hazard risks of flood and drought in rural areas. E-mail address: g_g_smitty@hotmail.com

MOHAMMED SALIM UDDIN is pursuing his PhD in natural resources and environmental management at the Natural Resources Institute, University of Manitoba, Winnipeg, where he earned his MNRM. He is assistant professor at the Department of Forestry and Environmental Sciences in the Shahjalal University of Science and Technology, Sylhet, Bangladesh. He has published in environmental management, forest management, and non-timber forest products. Currently he is working in natural disaster management with a special focus on resilience theory. E-mail address: salimuddin07@yahoo.com.